Name _____ Class _____

Skills Worksheet

Directed Reading

Section: From Genes to Proteins

In the space provided, write the letter of the description that best matches the term or phrase.

_____ 1. ribonucleic acid (RNA)

_____ 2. uracil

_____ 3. transcription

_____ 4. translation

_____ 5. gene expression

a. the entire process by which proteins are made

b. a molecule made of linked nucleotides

c. the process of reading instructions on an RNA molecule to put together the amino acids that make up a protein

d. the process of transferring a gene's instructions for making a protein to an RNA molecule

e. a nitrogen base used in RNA instead of the base thymine found in DNA

Complete each statement by underlining the correct term or phrase in the brackets.

6. Transcription begins when [RNA / RNA polymerase] binds to the gene's promoter.

7. RNA polymerase adds complementary [DNA / RNA] nucleotides as it "reads" the gene.

8. In eukaryotes, transcription takes place in the [nucleus / cytoplasm].

Read each question, and write your answer in the space provided.

9. What are two differences between transcription and DNA replication?

_____

_____

_____

_____

_____

10. What determines where on the DNA molecule transcription begins and where it ends?

_____

_____

Copyright © by Holt, Rinehart and Winston. All rights reserved.

Holt Biology — How Proteins Are Made

Name _____ Class _____ Date _____

Directed Reading continued

In the space provided, explain how the terms in each pair are related to each other.

11. RNA, messenger RNA

_____

_____

_____

12. codons, genetic code

_____

_____

_____

Study the following six steps in the synthesis of proteins. Determine the order in which the steps take place. Write the number of each step in the space provided.

_____ **13.** The codon in the vacant A site receives the tRNA molecule with the complementary anticodon. The tRNA carries the amino acid specified by the codon.

_____ **14.** Steps 2–5 are repeated until a stop codon is reached. The newly made protein is released into the cell.

_____ **15.** The tRNA at the P site detaches, leaves behind its amino acid, and moves away from the ribosome.

_____ **16.** Enzymes help form a peptide bond between the amino acids of adjacent tRNA molecules.

_____ **17.** The tRNA (with its protein chain) in the A site moves over to fill the empty P site. A new codon is present in the A site, ready to receive the next tRNA and its amino acid.

_____ **18.** An mRNA, two ribosomal subunits, and a tRNA carrying a modified form of the amino acid methionine bind together. The tRNA bonds to the "start" codon AUG.

Name _____ Class _____ Date _____

Skills Worksheet

Directed Reading

Section: Gene Regulation and Structure

Complete each statement by writing the correct term or phrase in the space provided.

1. To break down lactose, *Escherichia coli* need three different _____, each of which is coded for by a different gene.

2. The three genes are located next to each other, and all are controlled by the same _____ site.

3. The piece of DNA that overlaps the promoter site and serves as the on-off switch is called a(n) _____.

4. The group of genes that codes for enzymes involved in the same function, their promoter site, and the operator all function together as a(n) _____.

5. The operon that controls the metabolism of lactose is called the _____ _____.

6. A(n) _____ is a protein that binds to an operator and physically blocks RNA polymerase from binding to a promoter site.

Read each question, and write your answer in the space provided.

7. What are enhancers?

_____

_____

8. Why is there more opportunity for gene regulation in eukaryotic cells than in prokaryotic cells?

_____

_____

9. Why have no operons been found in eukaryotic cells?

_____

_____

Name _____ Class _____ Date _____

Directed Reading continued

10. When can gene regulation occur in eukaryotic cells?

   _____

   _____

   _____

11. What are introns and exons?

   _____

   _____

   _____

12. What happens to mRNA that includes introns?

   _____

   _____

   _____

13. What might be the evolutionary advantage of genes being interrupted by introns?

   _____

   _____

   _____

Complete each statement by underlining the correct term or phrase in the brackets.

14. Mutations can only be passed on to offspring if they occur in [gametes / body cells].

15. Mutations that change one or just a few nucleotides in a gene on a chromosome are called [random / point] mutations.

16. If a mutation causes a gene containing the nucleotide sequence ACA to become ACT, the mutation is called a [substitution / deletion] mutation.

17. If a mutation causes a sequence of nucleotides to change from ACGAGA to ACGGA, the mutation is called a(n) [insertion / deletion] mutation.

18. If a mutation causes a sequence of nucleotides to change from ACGAGA to ACGAGGA, the mutation is called a(n) [insertion / deletion] mutation.

Name _____ Class _____ Date _____

Skills Worksheet

Active Reading

Section: From Genes to Proteins

Read the passage below. Then answer the questions that follow.

Like DNA, **ribonucleic acid (RNA)** is a nucleic acid—a molecule made of nucleotides linked together. RNA differs from DNA in three ways. First, RNA consists of a single strand of nucleotides instead of the two strands found in DNA. Second, RNA nucleotides contain the five-carbon sugar ribose rather than the sugar deoxyribose found in DNA nucleotides. And third, RNA has a nitrogen base called **uracil**—abbreviated as *U*—instead of the base thymine (T) found in DNA. No thymine (T) bases are found in RNA. Like thymine, uracil is complementary to adenine whenever RNA base-pairs with another nucleic acid.

SKILL: RECOGNIZING SIMILARITIES AND DIFFERENCES

Read each question, and write your answer in the space provided.

1. In the spaces provided, write *D* if the statement is true of DNA. Write *R* if the statement is true of RNA. Write *B* if the statement is true of both DNA and RNA.

   _____ **a.** consists of a single strand of nucleotides

   _____ **b.** made of nucleotides linked together

   _____ **c.** contains deoxyribose

   _____ **d.** has the nitrogen base uracil

   _____ **e.** contains ribose

   _____ **f.** is a nucleic acid

   _____ **g.** consists of a double strand of nucleotides

   _____ **h.** contains a base that pairs with adenine

An analogy is a comparison. In the space provided, write the letter of the term or phrase that best completes the analogy.

   _____ **2.** RNA is to *U* as DNA is to
       a. *C*
       b. *G*
       c. *T*
       d. *A*

Copyright © by Holt, Rinehart and Winston. All rights reserved.

Holt Biology   How Proteins Are Made

Name _____ Class _____ Date _____

Skills Worksheet

Active Reading

Section: Gene Regulation and Structure

Read the passage below. Then answer the questions that follow.

A change in the DNA of a gene is called a mutation. The effects of a mutation vary, depending on whether it occurs in a gamete or in a body cell. Mutations in gametes can be passed on to offspring of the affected individual, but mutations in body cells affect only the individual in which they occur.

Mutations that move an entire gene to a new location are called *gene rearrangements*. Changes in a gene's position often disrupt the gene's function because the gene is exposed to new regulatory controls in its new location. This is something like moving to France and not being able to speak French.

Mutations that change a gene are called *gene alterations*. Gene alterations usually result in the placement of the wrong amino acid during protein assembly. This error can disrupt the protein's function. In a **point mutation**, a single nucleotide changes. In an *insertion* mutation, a sizable length of DNA is inserted into a gene. Insertions often result when mobile segments of DNA, called transposons, move randomly from one position to another on chromosomes. In a *deletion* mutation, segments of a gene are lost, often during meiosis.

SKILL: READING EFFECTIVELY

Read each question, and write your answer in the space provided.

1. What is a mutation?

_____

_____

2. A certain mutation is passed to offspring of the affected individual. What does this indicate about the type of cell in which the mutation originally occurred?

_____

_____

3. What is the difference between a gene rearrangement and a gene alteration?

_____

_____

Copyright © by Holt, Rinehart and Winston. All rights reserved.

Holt Biology   How Proteins Are Made

Name _____ Class _____ Date _____

Active Reading *continued*

4. What is an insertion?

_____

_____

_____

5. Why can a deletion have potentially catastrophic results?

_____

_____

_____

_____

In the space provided, write the letter of the phrase that best completes the statement.

_____ **6.** A mutation in a body cell is similar to a mutation in a gamete in that both involve
   a. offspring of the affected individual.
   b. a change in the DNA of a gene.
   c. addition of nucleotides.
   d. deletion of nucleotides.

Skills Worksheet
Vocabulary Review

Complete the crossword puzzle using the clues provided.

ACROSS

1. Like DNA, _____ acid (RNA) is a molecule made of nucleotides linked together.
2. RNA _____ is an enzyme involved in transcription.
3. The type of RNA that carries the instructions for making a protein from a gene to the site of translation is called _____ RNA.
4. The entire process by which proteins are made is called _____ expression.
6. process for transferring a gene's instructions for making a protein to an mRNA molecule
10. a three-nucleotide sequence on the mRNA that specifies an amino acid or "start" or "stop" signal
12. piece of DNA that serves as an on-off switch for transcription
14. long segment of nucleotides on a eukaryotic gene that has no coding information

7. a three-nucleotide sequence on a tRNA that is complementary to one of the codons of the genetic code
8. RNA molecules that are part of the structure of ribosomes are called _____ RNA.
9. RNA molecules that temporarily carry a specific amino acid on one end are called _____ RNA.
11. The _____ code specifies the amino acids and "start" and "stop" signals with their codon.
13. collective name for a group of genes involved in the same function, their promoter site, and their operator

DOWN

1. a protein that binds to an operator and inhibits transcription
5. portion of a eukaryotic gene that is translated
6. a process that puts together the amino acids that make up a protein

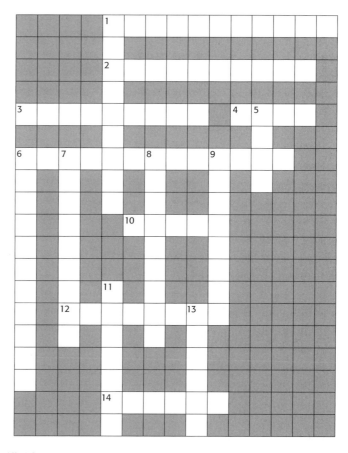

Copyright © by Holt, Rinehart and Winston. All rights reserved.

Holt Biology   How Proteins Are Made

Name _____ Class _____ Date _____

Skills Worksheet

Science Skills

Interpreting Tables

Use the table below to complete items 1–17.

Codons in mRNA					
First base	**Second base**				**Third base**
	U	**C**	**A**	**G**	
U	UUU ⎤ Phenylalanine UUC ⎦ UUA ⎤ Leucine UUG ⎦	UCU ⎤ UCC ⎥ Serine UCA ⎥ UCG ⎦	UAU ⎤ Tyrosine UAC ⎦ UAA ⎤ Stop UAG ⎦	UGU ⎤ Cysteine UGC ⎦ UGA – Stop UGG – Tryptophan	U C A G
C	CUU ⎤ CUC ⎥ Leucine CUA ⎥ CUG ⎦	CCU ⎤ CCC ⎥ Proline CCA ⎥ CCG ⎦	CAU ⎤ Histidine CAC ⎦ CAA ⎤ Glutamine CAG ⎦	CGU ⎤ CGC ⎥ Arginine CGA ⎥ CGG ⎦	U C A G
A	AUU ⎤ AUC ⎥ Isoleucine AUA ⎦ AUG – Start	ACU ⎤ ACC ⎥ Threonine ACA ⎥ ACG ⎦	AAU ⎤ Asparagine AAC ⎦ AAA ⎤ Lysine AAG ⎦	AGU ⎤ Serine AGC ⎦ AGA ⎤ Arginine AGG ⎦	U C A G
G	GUU ⎤ GUC ⎥ Valine GUA ⎥ GUG ⎦	GCU ⎤ GCC ⎥ Alanine GCA ⎥ GCG ⎦	GAU ⎤ Aspartic acid GAC ⎦ GAA ⎤ Glutamic acid GAG ⎦	GGU ⎤ GGC ⎥ Glycine GGA ⎥ GGG ⎦	U C A G

Complete the table below showing sequences of DNA, mRNA codons, anticodons, and corresponding amino acids. Use the list of mRNA codons in the table above to assist you in completing this exercise. Remember that the genetic code is based on mRNA codons.

Decoding DNA				
DNA	1. _____	2. _____	GAT	3. _____
mRNA codon	4. _____	5. _____	6. _____	UAU
Anticodon	7. _____	UUC	8. _____	9. _____
Amino acid	Tryptophan	10. _____	11. _____	12. _____

Copyright © by Holt, Rinehart and Winston. All rights reserved.

Holt Biology — How Proteins Are Made

Science Skills continued

Determine how the mutations below will affect each amino acid sequence. Use the mRNA codons in the table on the previous page to complete items a–d below. In the space provided, write the names of the amino acids that correspond to each mRNA sequence and mutation given. An example is provided for you.

Example:

mRNA sequence:	UGU-CCG	cysteine-proline
mutation sequence:	UGC-CGC	cysteine-arginine

13. mRNA sequence: GAA-CGU _____

    mutation sequence: GAU-CGU _____

14. mRNA sequence: AUC-UGC _____

    mutation sequence: AUC-UGG _____

15. mRNA sequence: UGU-CCU-CCU _____

    mutation sequence: UGU-UUC-CCU _____

16. mRNA sequence: GGG-UUA-ACC _____

    mutation sequence: GGU-UAA _____

17. What kind of mutation occurred to the mRNA sequence in item 16 above? Explain.

_____

_____

_____

_____

_____

Holt Biology — How Proteins Are Made

Name _____ Class _____ Date _____

Skills Worksheet

Concept Mapping

Using the terms and phrases provided below, complete the concept map showing the principles of protein synthesis.

    amino acids   RNA   transcription

    genes   RNA polymerase   translation

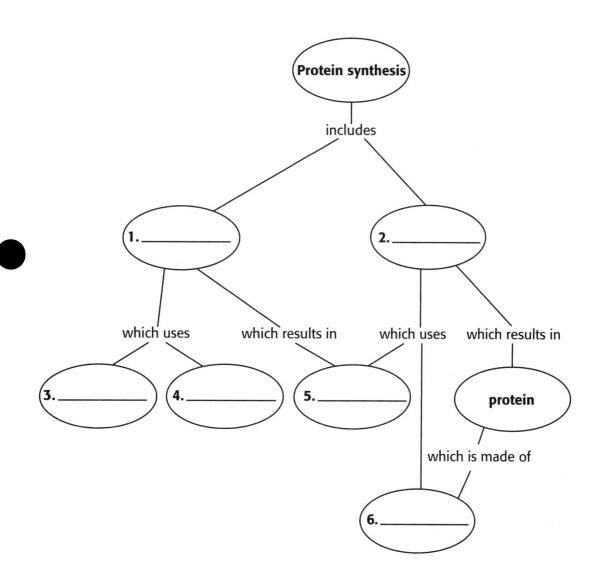

Holt Biology   How Proteins Are Made

Name _____ Class _____ Date _____

Skills Worksheet

Critical Thinking

Work-Alikes

In the space provided, write the letter of the term or phrase that best describes how each numbered item functions.

_____ 1. mRNA

_____ 2. transcription

_____ 3. operator

_____ 4. RNA polymerase

_____ 5. point mutation

_____ 6. translation

a. electrical switch
b. computer disk with data
c. e-mail
d. interpreter translating from one language to another
e. substitute teacher
f. train on a track

Cause and Effect

In the space provided, write the letter of the term or phrase that best matches each cause or effect given below.

Cause	Effect	
7. anticodons bond to codons	_____	a. all life-forms have the same evolutionary ancestors
8. codons are the same in all organisms	_____	b. genetic flexibility
9. _____	translation begins	c. amino acids are brought together to form proteins
10. _____	*lac* operon turned on	d. mRNA leaves the nucleus and enters the cytoplasm
11. introns in DNA	_____	e. few thousand exons
12. _____	thousands of proteins in human cells	f. presence of lactose

Copyright © by Holt, Rinehart and Winston. All rights reserved.

Holt Biology — How Proteins Are Made

Linkages

In the spaces provided, write the letters of the two terms or phrases that are linked together by the term or phrase in the middle. The choices can be placed in any order.

13. _____ RNA transcription _____
14. _____ rRNA _____
15. _____ ribosome _____
16. _____ translation _____
17. _____ binds to operator _____
18. _____ repressor of *lac* operon changes shape _____

a. operon switched off
b. mRNA sequence
c. repressor
d. mRNA
e. amino acid sequence
f. RNA translation
g. P site
h. DNA replication
i. lactose
j. operon switched on
k. A site
l. tRNA

Analogies

An analogy is a relationship between two pairs of terms or phrases written as a : b :: c : d. The symbol : is read as "is to," and the symbol :: is read as "as." In the space provided, write the letter of the pair of terms or phrases that best completes the analogy shown.

_____ 19. DNA : thymine ::
   a. thymine : RNA
   b. DNA : uracil
   c. DNA : ribose
   d. RNA : uracil

_____ 20. mRNA : transcription ::
   a. DNA : transcription
   b. rRNA : translation
   c. protein : translation
   d. DNA : translation

_____ 21. both DNA strands : replication ::
   a. both DNA strands : transcription
   b. one DNA strand : replication
   c. one DNA strand : transcription
   d. one DNA strand : translation

_____ 22. transcription in prokaryotic cells : cytoplasm ::
   a. uracil : DNA
   b. ribose : RNA
   c. transcription in eukaryotes : cytoplasm
   d. transcription in eukaryotes : nucleus

_____ 23. DNA promoter sequence : transcription ::
   a. brake : car
   b. bread : peanut butter
   c. stop sign : traffic
   d. track starter's pistol : track event

Name _____ Class _____ Date _____

Skills Worksheet

Test Prep Pretest

Complete each statement by writing the correct term or phrase in the space provided.

1. Instead of the base thymine found in DNA, RNA has a base called _____.

2. Transcription begins when an enzyme called _____ _____ binds to the beginning of a gene on a region of DNA called a promoter.

3. The instructions for building a protein are written as a series of three-nucleotide sequences called _____.

4. During translation, the area of the ribosome called the _____ site receives the next tRNA molecule.

5. Because of its position on the operon, the _____ is able to control RNA polymerase's access to the structural genes.

6. The *lac* operon is switched off when a protein called a(n) _____ is bound to the operator.

7. In eukaryotic gene regulation, proteins called _____ _____ help arrange RNA polymerases in the correct position on the promoter.

8. In eukaryotes, long segments of nucleotides with no coding information are called _____.

9. In eukaryotes, the portions of a gene that are actually translated into proteins are called _____.

10. Insertions, deletions and point mutations are types of _____ _____.

Copyright © by Holt, Rinehart and Winston. All rights reserved.

Holt Biology — How Proteins Are Made

Name _____ Class _____ Date _____

| Test Prep Pretest *continued*

Questions 11–13 refer to the figure below.

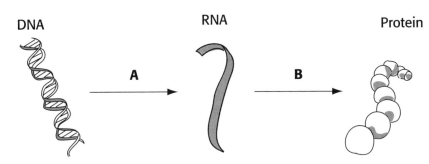

11. The processing of information from DNA into proteins, as shown above, is referred to as _____ _____ .

12. Stage *A* is called _____ .

13. Stage *B* is called _____ .

In the space provided, write the letter of the term or phrase that best completes each statement or best answers each question.

_____ 14. In what kinds of cells do mutations occur?
   a. body cells
   b. gametes
   c. reproductive cells
   d. All of the above

_____ 15. A mutation that moves a gene to a new location is called a(n)
   a. point mutation.
   b. insertion.
   c. transposon.
   d. deletion.

_____ 16. Which of the following represents the codons that correspond to this segment of DNA: TATCAGGAT?
   a. AUA—GUC—CUA
   b. ATA—GTC—CTA
   c. AUAGU—CCUA
   d. ACA—CUC—GUA

_____ 17. Which of the following are the anticodons that correspond to the mRNA codons CAG—ACU—UUU?
   a. GTC—TGA—AAA
   b. GUC—UGA—AAA
   c. glutamine—threonine—phenylalanine
   d. GAC—UCA—AAA

_____ 18. Because the genetic code is the same in all organisms, it appears that
   a. the genetic code evolved more than once.
   b. the codon GUC codes for different proteins in different organisms.
   c. thymine will soon replace uracil in RNA.
   d. all life-forms have a common ancestor.

Name _____ Class _____ Date _____

Test Prep Pretest continued

Read each question, and write your answer in the space provided.

19. Explain how RNA differs from DNA.

20. Summarize the process of translation.

21. Describe the functions of RNA.

22. What is the *lac* operon?

23. Explain why gene regulation in eukaryotic cells is more complex than in prokaryotic cells.

Name _____ Class _____ Date _____

Test Prep Pretest continued

24. Why do scientists think that introns and exons contribute to evolutionary flexibility?

_____

_____

_____

_____

25. Describe the three ways that mutation can alter genetic material.

_____

_____

_____

_____

_____

Name _____ Class _____ Date _____

Assessment
Quiz

Section: From Genes to Proteins

In the space provided, write the letter of the term or phrase that best completes each statement or best answers each question.

_____ 1. During transcription, the genetic information for making a protein is rewritten as a molecule of
   a. messenger RNA.
   b. ribosomal RNA.
   c. transfer RNA.
   d. translation RNA.

_____ 2. All organisms have a genetic code made of
   a. two-nucleotide sequences.
   b. three-nucleotide sequences.
   c. four-nucleotide sequences.
   d. five-nucleotide sequences.

_____ 3. In a cell, the equipment for translation is located in the
   a. cytoplasm.
   b. nucleus.
   c. plasma membrane.
   d. centrioles.

_____ 4. Like DNA, RNA contains which of the following?
   a. phosphate
   b. uracil
   c. thymine
   d. deoxyribose

_____ 5. In eukaryotes, translation begins when
   a. the A site becomes vacant.
   b. tRNA detaches from the P site.
   c. a stop codon is reached.
   d. mRNA leaves the nucleus.

In the space provided, write the letter of the description that best matches the term or phrase.

_____ 6. RNA

_____ 7. mRNA

_____ 8. RNA polymerase

_____ 9. tRNA

_____ 10. DNA

a. enzyme that adds and links complementary RNA nucleotides during transcription

b. helps in the synthesis of proteins by carrying amino acids

c. single strand of nucleotides containing ribose and uracil

d. double strand of nucleotides containing deoxyribose and thymine

e. delivers the information needed to make a protein to the site of translation

Copyright © by Holt, Rinehart and Winston. All rights reserved.
Holt Biology   How Proteins Are Made

Name _____ Class _____ Date _____

Assessment

Quiz

Section: Gene Regulation and Structure

In the space provided, write the letter of the term or phrase that best completes each statement or best answers each question.

_____ 1. Gene regulation is necessary in living organisms
   a. so that the repressor will never bind to the operator.
   b. to allow RNA polymerase continuous access to genes.
   c. to avoid wasting their energy and resources on producing proteins that are not needed or are already available.
   d. to ensure that the operon is always in the "on" mode.

_____ 2. The *lac* operon enables a bacterium to build the proteins needed for lactose metabolism only when
   a. glucose is present.
   b. tryptophan is present.
   c. galactose is present.
   d. lactose is present.

_____ 3. Which of the following is NOT true about gene regulation in eukaryotic cells?
   a. Gene regulation in eukaryotes is more complex than in prokaryotes.
   b. Operons play a major role in eukaryote gene regulation.
   c. Gene regulation can occur before, during, or after transcription.
   d. Gene regulation can occur after translation.

_____ 4. Point mutations occur when
   a. one nucleotide is replaced with a different nucleotide.
   b. a gene's location changes.
   c. long segments of a gene are lost.
   d. gametes are forming during meiosis.

_____ 5. The *lac* operon turns "off" when
   a. glucose is absent.
   b. lactose is absent.
   c. RNA polymerase is absent.
   d. lactose is present.

In the space provided, write the letter of the description that best matches the term or phrase.

_____ 6. intron

_____ 7. gene alteration

_____ 8. exon

_____ 9. repressor

_____ 10. enhancer

a. long segments of eukaryotic DNA that have no coding information
b. includes the substitution, insertion, and deletion of one or more nucleotides
c. sequence of DNA that can be bound to a transcription factor
d. can bind to an operator, which stops transcription
e. portions of a eukaryotic gene that are translated

Copyright © by Holt, Rinehart and Winston. All rights reserved.
Holt Biology — How Proteins Are Made

Name _____ Class _____ Date _____

Assessment

Chapter Test

How Proteins Are Made

In the space provided, write the letter of the term or phrase that best completes each statement or best answers each question.

_____ 1. RNA differs from DNA in that RNA
   a. is single-stranded.
   b. contains a different sugar.
   c. contains uracil.
   d. All of the above

_____ 2. After mRNA has been transcribed in eukaryotes,
   a. its introns are cut out.
   b. its exons are joined together.
   c. it leaves the nucleus through pores.
   d. All of the above

_____ 3. The enzyme that adds and links complementary RNA nucleotides during transcription is called
   a. RNA polymerase.
   b. DNA polymerase.
   c. an operon.
   d. an enhancer.

_____ 4. At the beginning of translation, the first tRNA molecule
   a. binds to the ribosome's A site.
   b. attaches directly to the DNA codon.
   c. connects an amino acid to its anticodon.
   d. attaches to the P site of the ribosome.

_____ 5. An operon is composed of which of the following?
   a. a group of proteins, their promoter site, and their operator
   b. a group of genes, their operator, and RNA polymerase
   c. a group of genes, their promoter site, and their operator
   d. an enhancer, an operator, and RNA polymerase

_____ 6. Regulatory proteins in eukaryotes that are involved in controlling the onset of transcription are called
   a. repressors.
   b. transcription factors.
   c. operators.
   d. enhancers.

_____ 7. If a deletion mutation eliminated all of the guanine bases from the codon sequence GAT-CGC-CAA-TAG, the altered sequence would read
   a. ATC-TCA-ATA.
   b. ATC-CCA-ATA.
   c. ACC-CAA-ATA.
   d. AAT-CCA-TAC.

Copyright © by Holt, Rinehart and Winston. All rights reserved.

Holt Biology · How Proteins Are Made

Name _____ Class _____ Date _____

Chapter Test continued

Questions 8–10 refer to the mRNA sequence CUC-AAG-UGC-UUC and the table below, which lists mRNA codons.

		Codons in mRNA			
First base		**Second base**			**Third base**
	U	**C**	**A**	**G**	
U	UUU, UUC Phenylalanine UUA, UUG Leucine	UCU, UCC, UCA, UCG Serine	UAU, UAC Tyrosine UAA, UAG Stop	UGU, UGC Cysteine UGA – Stop UGG – Tryptophan	U C A G
C	CUU, CUC, CUA, CUG Leucine	CCU, CCC, CCA, CCG Proline	CAU, CAC Histidine CAA, CAG Glutamine	CGU, CGC, CGA, CGG Arginine	U C A G
A	AUU, AUC, AUA Isoleucine AUG – Start	ACU, ACC, ACA, ACG Threonine	AAU, AAC Asparagine AAA, AAG Lysine	AGU, AGC Serine AGA, AGG Arginine	U C A G
G	GUU, GUC, GUA, GUG Valine	GCU, GCC, GCA, GCG Alanine	GAU, GAC Aspartic acid GAA, GAG Glutamic acid	GGU, GGC, GGA, GGG Glycine	U C A G

_____ 8. Which of the following would represent the sequence of DNA from which the mRNA sequence was made?
  a. CUC-AAG-UGC-UUC
  b. GAG-UUC-ACG-AAG
  c. GAG-TTC-ACG-AAG
  d. AGA-CCT-GTA-GGA

_____ 9. The anticodons for the codons in the mRNA sequence above are
  a. GAG-UUC-ACG-AAG.
  b. GAG-TTC-ACG-AAG.
  c. CUC-GAA-CGU-CUU.
  d. CUU-CGU-GAA-CUC.

_____ 10. Which of the following represents the portion of the protein molecule coded for by the mRNA sequence above?
  a. serine-tyrosine-arginine-glycine
  b. valine-aspartic acid-proline-histidine
  c. leucine-lysine-cysteine-phenylalanine
  d. glutamic acid-phenylalanine-threonine-lysine

Name _____ Class _____ Date _____

Chapter Test *continued*

_____ 11. Which of the following observations supports the conclusion that all life-forms have a common evolutionary ancestor with a single genetic code?
   a. Some cell organelles do not read stop codons.
   b. The genetic code is the same in all organisms.
   c. Animals of different species are not able to produce offspring together.
   d. Each codon codes for a different protein in different organisms.

_____ 12. Eukaryotic genes
   a. contain operons.
   b. have long, unbroken stretches of DNA called spliceosomes.
   c. are long, unbroken stretches of nucleotides that all code for a single protein.
   d. are interrupted by non-coding segments called introns.

In the space provided, write the letter of the description that best matches the term or phrase.

_____ 13. transcription

_____ 14. translation

_____ 15. gene alteration

_____ 16. RNA polymerase

_____ 17. codon

_____ 18. transcription factor

_____ 19. anticodon

_____ 20. repressor

a. can cause the placement of the wrong amino acid during protein assembly

b. three-nucleotide sequence in tRNA

c. enzyme that builds an mRNA from DNA

d. can turn the lac operon "on" and "off"

e. three-nucleotide sequences in mRNA

f. transferring genetic information from a gene to mRNA

g. initiate transcription by binding to enhancers and RNA polymerases

h. putting together the amino acids that make up a protein

Name _____ Class _____ Date _____

Assessment

Chapter Test

How Proteins Are Made

Complete each statement by writing the correct term or phrase in the space provided.

1. The information contained in a molecule of mRNA is used to make proteins during the process of _____ .

2. Nucleotide sequences of tRNA that are complementary to codons on mRNA are called _____ .

3. Nucleotides that make up RNA contain the nitrogen bases adenine, guanine, cytosine, or _____ .

4. With few exceptions, all organisms use the same _____ _____ .

5. When there is no lactose present in a bacterial cell, a(n) _____ turns the *lac* operon _____ .

6. A repressor can physically block _____ _____ from binding to a promoter site on the *lac* operon.

7. In _____ , gene regulation can occur before, during, and after transcription.

8. The sequence of nucleotides in RNA is _____ to those that make up a gene.

In the space provided, write the letter of the term or phrase that best completes each statement or best answers each question.

_____ 9. The piece of DNA that overlaps the promoter and severs as the on-off switch in operons is called a(n)
   a. promoter site.
   b. operator.
   c. enhancer.
   d. repressor.

_____ 10. An operon includes which of the following?
   a. a promoter site
   b. an operator
   c. a group of genes that code for enzymes involved in the same function
   d. All of the above

Copyright © by Holt, Rinehart and Winston. All rights reserved.

Holt Biology

Name _____ Class _____ Date _____

Chapter Test continued

_____ 11. A mutation in which only one nucleotide is changed in a gene is called a(n)
   a. insertion.
   b. deletion.
   c. point mutation.
   d. transposition.

_____ 12. In eukaryotes, gene expression is regulated by
   a. operons.
   b. operators.
   c. transcription factors.
   d. All of the above

_____ 13. Which of the following occurs in eukaryotes?
   a. Exons are cut out after transcription.
   b. Only exons are translated.
   c. Only introns are transcribed.
   d. Introns are joined together and then translated.

_____ 14. The function of tRNA is to
   a. synthesize DNA.
   b. synthesize mRNA.
   c. form ribosomes.
   d. transfer amino acids to ribosomes.

_____ 15. In the absence of lactose, a repressor molecule binds to
   a. the operator.
   b. another repressor.
   c. the enhancer.
   d. an exon.

_____ 16. Which of the following is an advantage of regulating gene expression?
   a. The organism conserves energy.
   b. The organism can prevent mutations.
   c. Most genes are never transcribed.
   d. Fewer genes have to be passed on to the next generation.

_____ 17. The function of mRNA is to
   a. synthesize DNA.
   b. carry information from genes to ribosomes.
   c. form ribosomes.
   d. transfer amino acids to ribosomes.

_____ 18. Translation occurs
   a. in the nucleus.
   b. only in prokaryotes.
   c. on ribosomes.
   d. on a DNA template.

Name _____ Class _____ Date _____

Chapter Test *continued*

Read each question, and write your answer in the space provided.

19. What is the difference between transcription and translation?

_____
_____
_____
_____

20. Explain the evolutionary significance of the fact that the base sequence GUC codes for the amino acid valine in so many different species.

_____
_____
_____
_____

21. How is RNA polymerase affected by the presence of lactose in bacterial cells?

_____
_____
_____
_____

22. What are introns?

_____
_____
_____
_____

23. What is a gene alteration? Give two examples.

_____
_____
_____
_____

Name _____ Class _____ Date _____

Chapter Test continued

24. Describe how eukaryotic genes are organized.

_____

_____

_____

_____

_____

25. Summarize the role of transcription factors in regulating eukaryotic gene expression.

_____

_____

_____

_____

_____

_____

_____

Name _____ Class _____ Date _____

Quick Lab

DATASHEET FOR IN-TEXT LAB

Modeling Transcription

You can use paper and pens to model the process of transcription.

MATERIALS
- paper
- scissors
- pens or pencils (two colors)
- tape

Procedure

1. Cut a sheet of paper into 36 squares, each about 2.5 × 2.5 cm (1 × 1 in.) in size.

2. To make one side of your DNA model, line up 12 squares in a column. Using one color, randomly label each square with one of the following letters: A, C, G, or T. Each square represents a DNA nucleotide. Use tape to keep the squares in a column.

3. To make the second side of your DNA model, line up 12 squares next to the first column. Use the same color you used in step 2 to label each square with the complementary DNA nucleotide. Tape the squares together in a column.

4. Separate the two columns. The remaining 12 squares represent RNA nucleotides. Use a different color to "transcribe" one of the DNA strands.

Analysis

1. **Propose** a reason for using different colors for the DNA and RNA "nucleotides."

   _____
   _____
   _____

2. **Predict** how a change in the sequence of nucleotides in a DNA molecule would affect the mRNA transcribed from the DNA molecule.

   _____
   _____
   _____

Copyright © by Holt, Rinehart and Winston. All rights reserved.

Name _____ Class _____ Date _____

Modeling Transcription *continued*

3. Critical Thinking
  Applying Information Use your model to test your prediction. Describe your results.

_____

_____

_____

_____

Name _____ Class _____ Date _____

Data Lab

Decoding the Genetic Code

DATASHEET FOR IN-TEXT LAB

Background

Keratin is one of the proteins in hair. The gene for keratin is transcribed and translated by certain skin cells. The series of letters below represents the sequence of nucleotides in a portion of an mRNA molecule transcribed from the gene for keratin. This mRNA strand and the genetic code in the table below can be used to determine some of the amino acids in keratin.

mRNA Segment

U C U C G U G A A U U U U C C

First base	Second base				Third base
	U	**C**	**A**	**G**	
U	UUU ⎤ Phenylalanine UUC ⎦ UUA ⎤ Leucine UUG ⎦	UCU ⎤ UCC ⎥ Serine UCA ⎥ UCG ⎦	UAU ⎤ Tyrosine UAC ⎦ UAA ⎤ Stop UAG ⎦	UGU ⎤ Cysteine UGC ⎦ UGA – Stop UGG – Tryptophan	U C A G
C	CUU ⎤ CUC ⎥ Leucine CUA ⎥ CUG ⎦	CCU ⎤ CCC ⎥ Proline CCA ⎥ CCG ⎦	CAU ⎤ Histidine CAC ⎦ CAA ⎤ Glutamine CAG ⎦	CGU ⎤ CGC ⎥ Arginine CGA ⎥ CGG ⎦	U C A G
A	AUU ⎤ AUC ⎥ Isoleucine AUA ⎦ AUG – Start	ACU ⎤ ACC ⎥ Threonine ACA ⎥ ACG ⎦	AAU ⎤ Asparagine AAC ⎦ AAA ⎤ Lysine AAG ⎦	AGU ⎤ Serine AGC ⎦ AGA ⎤ Arginine AGG ⎦	U C A G
G	GUU ⎤ GUC ⎥ Valine GUA ⎥ GUG ⎦	GCU ⎤ GCC ⎥ Alanine GCA ⎥ GCG ⎦	GAU ⎤ Aspartic acid GAC ⎦ GAA ⎤ Glutamic acid GAG ⎦	GGU ⎤ GGC ⎥ Glycine GGA ⎥ GGG ⎦	U C A G

Analysis

1. Determine the sequence of amino acids that will result from the translation of the segment of mRNA above.

_____

2. Determine the anticodon of each tRNA molecule that will bind to this mRNA segment.

_____

Name _____ Class _____ Date _____

Decoding the Genetic Code continued

3. Critical Thinking
Recognizing Patterns Determine the sequence of nucleotides in the segment of DNA from which the mRNA strand on the previous page was transcribed.

_____

4. Critical Thinking
Recognizing Patterns Determine the sequence of nucleotides in the segment of DNA that is complementary to the DNA segment described in item 3.

_____

Name _____ Class _____ Date _____

Quick Lab

DATASHEET FOR IN-TEXT LAB

Modeling Introns and Exons

You can use masking tape to represent introns and exons.

MATERIALS
- masking tape
- pens or pencils (two colors)
- metric ruler
- scissors

Procedure

1. Place a 15–20 cm strip of masking tape on your desk. The tape represents a gene.

2. Use two colors to write the words *appropriately joined* on the tape exactly as shown in the example. Space the letters so that they take up the entire length of the strip of tape. The segments in one color represent introns; those in the other color represent exons.

> appropriately joined

3. Lift the tape. Working from left to right, cut apart the groups of letters written in the same color. Stick the pieces of tape to your desk as you cut them, making two strips according to color and joining the pieces in their original order.

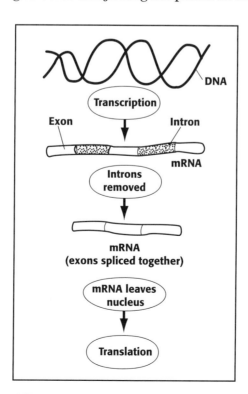

Copyright © by Holt, Rinehart and Winston. All rights reserved.

Holt Biology — How Proteins Are Made

Name _____ Class _____ Date _____

Modeling Introns and Exons *continued*

Analysis

1. **Determine** from the resulting two strips which strip is made of "introns" and which is made of "exons."

   _____

   _____

2. **Critical Thinking**
   Predicting Outcomes Predict what might happen to a protein if an intron were not removed.

   _____

   _____

   _____

   _____

   _____

Name _____ Class _____ Date _____

Exploration Lab

DATASHEET FOR IN-TEXT LAB

Modeling Protein Synthesis

SKILLS
- Modeling
- Using scientific methods

OBJECTIVES
- **Compare** and **Contrast** the structure and function of DNA and RNA.
- **Model** protein synthesis.
- **Demonstrate** how a mutation can affect a protein.

MATERIALS
- masking tape
- plastic soda-straw pieces of one color
- plastic soda-straw pieces of a different color
- paper clips
- pushpins of five different colors
- marking pens of the same colors as the pushpins
- 3 × 5 in. note cards
- oval-shaped card
- transparent tape

Before You Begin

The nature of a **protein** is determined by the sequence of amino acids in its structure. During **protein synthesis**, the sequence of nitrogen bases in an **mRNA** molecule is used to assemble **amino acids** into a protein chain.

A **mutation** is a change in the nitrogen-base sequence of DNA. Many mutations lead to altered or defective proteins. For example, the genetic blood disorder **sickle cell anemia** is caused by a mutation in the gene for **hemoglobin**.

In this lab, you will build models that will help you understand how protein synthesis occurs. You can also use the models to explore how a mutation affects a protein.

1. Write a definition for each boldface term in the paragraph above and for each of the following terms: transcription, translation, tRNA, ribosome, codon, anticodon. Use a separate sheet of paper.

Copyright © by Holt, Rinehart and Winston. All rights reserved.

Holt Biology

How Proteins Are Made

Name _____ Class _____ Date _____

Modeling Protein Synthesis continued

2. Describe three differences between DNA and RNA.

_____

_____

_____

_____

3. Based on the objectives for this lab, write a question you would like to explore about protein synthesis.

_____

_____

Procedure

PART A: DESIGN A MODEL

1. Work with the members of your lab group to design models of DNA, RNA, and a cell. Use the materials listed for this lab.

   > **You Choose**
   > As you design your models, decide the following:
   > a. what question you will explore
   > b. how to represent DNA nucleotides
   > c. how to represent RNA nucleotides
   > d. how to represent five different nitrogen bases
   > e. how to link (bond) nucleotides together
   > f. how to represent tRNA molecules with amino acids
   > g. how to represent the locations of DNA and ribosomes

2. Write out the plan for building your models. Have your teacher approve the plan before you begin building the models.

_____

_____

_____

_____

Name _____ Class _____ Date _____

Modeling Protein Synthesis continued

3. Build the models your group designed. **CAUTION: Sharp or pointed objects may cause injury. Handle pushpins carefully.** Start your model of DNA with a strand of nucleotides that has the following sequence of nitrogen bases: TTTGGTCTCCTC.

PART B: MODEL PROTEIN SYNTHESIS

4. Use your models to demonstrate how transcription and translation occur. Draw and label the steps of each process on a separate sheet of paper.

5. Use your models to explore one of the questions written for step 3 of **Before You Begin.**

_____
_____
_____
_____

PART C: TEST HYPOTHESIS

Answer each of the following questions by writing a hypothesis. Use your models to test each hypothesis, and describe your results.

6. The DNA model you built for step 3 represents a portion of a gene for hemoglobin. Sickle cell anemia results from the substitution of an A for the T in the third codon of the nitrogen-base sequence given in step 3. How will this substitution affect a hemoglobin molecule?

_____
_____

7. The addition of a nucleotide to a strand of DNA is a type of mutation called an *insertion*. What happens when an insertion occurs in the first codon in a DNA strand, before the DNA strand is transcribed?

_____
_____

PART D: CLEANUP AND DISPOSAL

8. Dispose of damaged pushpins in the designated waste container.

9. Clean up your work area and all lab equipment. Return lab equipment to its proper place. Wash your hands thoroughly before you leave the lab and after you finish all work.

Name _____ Class _____ Date _____

Modeling Protein Synthesis *continued*

Analyze and Conclude

1. **Comparing Structures** How did the nitrogen-base sequence of the mRNA you made compare with that of the DNA it was transcribed from?

   _____

   _____

2. **Recognizing Relationships** How is the nitrogen-base sequence of a gene related to the structure of a protein?

   _____

   _____

   _____

3. **Recognizing Patterns** What is the relationship between the anticodon of a tRNA and the amino acid the tRNA carries?

   _____

   _____

4. **Drawing Conclusions** How does a mutation in the gene for a protein affect the protein?

   _____

   _____

   _____

5. **Further Inquiry** Write a new question about protein synthesis that could be explored with your model.

   _____

   _____

   _____

Name _____ Class _____ Date _____

Exploration Lab

DNA Whodunit

BIOTECHNOLOGY

In the early 1970s, scientists discovered that some bacteria have enzymes that are able to cut up DNA in a sequence-specific manner. These enzymes, now called *restriction enzymes*, recognize and bind to a specific short sequence of DNA, and then cut the DNA at specific sites within that sequence. Biologists found that they could use restriction enzymes to manipulate DNA. This ability formed the foundation for much of the biotechnology that exists today.

DNA fingerprinting is one important use of biotechnology. With the exception of identical twins, no two people have the same DNA sequence. Because each person has a DNA profile that is as unique as his or her fingerprints, DNA fingerprinting can be used to compare the DNA of different individuals.

In the first step of DNA fingerprinting, known and unknown samples are obtained and then digested, or cut into small fragments, by the same restriction enzyme. These short fragments are called *restriction fragment length polymorphisms (RFLPs)*. The next step in DNA fingerprinting is to separate the RFLPs by size. This is done with a technique called *gel electrophoresis*. The DNA is placed on a jellylike slab called a gel, and the gel is exposed to an electrical current. DNA has a negative electrical charge, so the RFLPs are attracted to the positive pole when an electric current is applied. Smaller fragments travel farther through the gel than longer ones. The length of a given DNA fragment can be determined by comparing its mobility on the gel with that of a sample containing DNA fragments of known sizes. The resulting pattern is unique for each individual.

In this lab, you will model experimental procedures involved in DNA fingerprinting and use your results to identify a hypothetical murderer.

OBJECTIVES

Use pop beads to model restriction enzyme digestion and agarose gel electrophoresis.

Evaluate the results of a model restriction enzyme digestion (DNA fingerprint).

Identify a hypothetical murderer by analyzing the simulated DNA fingerprints of suspects and DNA samples collected at the scene of the crime.

MATERIALS

- pop beads, blue (cytosine) (15)
- paper gel electrophoresis lane
- paper, legal size (8.5 × 14 in.)
- plastic connectors (hydrogen bonds) (30)
- pop beads, green (guanine) (15)
- pop beads, orange (thymine) (15)
- pop beads, red (phosphate) (60)
- pop beads, white, 5-hole (deoxyribose) (60)
- pop beads, yellow (adenine) (15)
- restriction enzyme card Jan I
- restriction enzyme card Ward II
- ruler

Name _____ Class _____ Date _____

DNA Whodunit *continued*

Procedure

1. Read the following scenario.

   The police are investigating a murder. Blood stains of two different types were found at the murder scene. Based on other forensic evidence, the police have reason to believe that the murderer was wounded at the time of the murder. The police currently have five suspects for the murder. You have been provided with the DNA from a blood sample of one of the five suspects, or the DNA from one of the two blood stains found at the crime scene.

2. Assemble the DNA you were assigned with pop beads, using the DNA strip given to your group as a blueprint. Use **Figure 1** to guide you in your assembly of your DNA "molecule." Be sure to assemble the beads in the precise pattern indicated, or your results will be incorrect. The assembled chain represents your subject's DNA.

FIGURE 1 PATTERN FOR ASSEMBLY OF POP BEADS

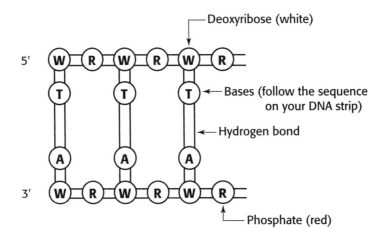

3. Place the DNA "molecule" you have just assembled on your work area so that the 5' end is on the top left side, as shown below. Be sure that the three orange beads (thymine) are in the following position on your work area:

   ```
           5'                      3'
         TTT, etc..................G
         AAA, etc.................C
           3'                      5'
   ```

Note: From this point on, it is important to keep the beads in this orientation. Do not allow the chain to be turned upside down or rotated. The 5' TTT end should always be on the top left of the molecule. If your chain is accidentally turned upside down, refer to your DNA strip to obtain the correct orientation.

Copyright © by Holt, Rinehart and Winston. All rights reserved.

Holt Biology — How Proteins Are Made

Name _____ Class _____ Date _____

DNA Whodunit *continued*

4. Use the model restriction enzymes Jan I and Ward II to chop up the DNA. Look at your two Restriction Enzyme Cards; they look like the cards in **Figure 2.** These "enzymes" will make cuts in the DNA in the manner indicated by the dotted lines.

FIGURE 2 RESTRICTION ENZYME CARDS

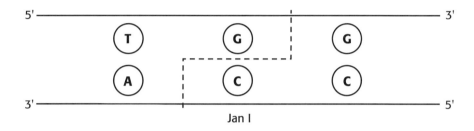
Jan I

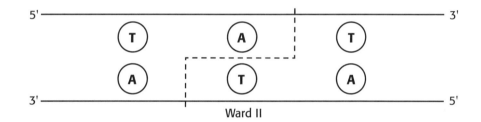
Ward II

5. Place Restriction Enzyme Card Jan I on top of the left side of the DNA chain so that its label is right side up.

6. Move the card along the surface of the DNA until you match the precise sequence shown on the card. When you reach a sequence that matches the card, stop and break the beads apart in the manner indicated by the dotted lines.

7. Move the enzyme card until you reach the right end of the DNA. Double check the sequence with the enzyme card to ensure that you have made all the possible cuts.

8. Repeat the procedure on the remaining DNA fragments using the restriction enzyme card Ward II. Be sure to keep the DNA fragments in the orientation described above (5' orange thymine beads on the top left) throughout this exercise. In reality, the fragments created in steps 6 and 7 might be thousands of base pairs long.

9. Create a gel electrophoresis area out of a legal size (8.5 by 14 in.) sheet of paper. On the left side of the paper, use a ruler to mark one inch increments from the bottom of the paper to the top of the paper. Starting at the bottom mark, label each mark from "0" (for the bottom mark) through '24" (for the top mark).

Copyright © by Holt, Rinehart and Winston. All rights reserved.

Holt Biology — How Proteins Are Made

DNA Whodunit continued

10. Write a plus sign (+) at the bottom of the page and a minus sign (−) at the top of the page. Label the *Y*-axis (left hand margin) "Length of RFLPs (number of nucleotides."

11. Place the RFLPs at the negative pole of the gel electrophoresis page, taking care to retain the proper 5' to 3' orientation. Remember, DNA has a negative electrical charge, so the RFLPs are attracted to the positive end of the gel/page when an electric current is applied.

12. Simulate separating the RFLPs by electrophoresis by sliding your RFLPs along the gel/page. Shorter fragments are lighter and move farther than longer fragments. To determine the final position of each RFLP, count the number of nucleotides on the longest side of each fragment. Place each measured RFLP next to its corresponding length marked on the gel/page.

13. In the nine gel electrophoresis lanes in **Figure 3,** sketch dark bands at the correct positions in the gel lane reserved for your sample. Also, record the position of your bands on the seven lanes your teacher has provided for class data (on the blackboard).

14. Obtain the banding patterns for each of the other DNA samples by copying them from the blackboard after each team has recorded their data.

FIGURE 3 GEL ELECTROPHORESIS LANES

Victim's Blood	Suspect 1	Suspect 2	Suspect 3	Suspect 4	Suspect 5	Crime Scene Sample 1	Crime Scene Sample 2
22	22	22	22	22	22	22	22
20	20	20	20	20	20	20	20
18	18	18	18	18	18	18	18
16	16	16	16	16	16	16	16
14	14	14	14	14	14	14	14
12	12	12	12	12	12	12	12
10	10	10	10	10	10	10	10
8	8	8	8	8	8	8	8
6	6	6	6	6	6	6	6
4	4	4	4	4	4	4	4
2	2	2	2	2	2	2	2

Name _____ Class _____ Date _____

DNA Whodunit *continued*

Analysis

1. Examining Data Are the RFLPs of the other DNA samples the same length as yours? Explain why or why not.

_____
_____
_____
_____
_____

2. Identifying Relationships Explain the role that restriction enzymes and gel electrophoresis play in DNA fingerprinting.

_____
_____
_____
_____

Conclusions

1. Drawing Conclusions Based on class data, which of the suspects is probably the murderer? Explain.

_____
_____
_____
_____
_____

2. Interpreting Information How did you show that the other sample found at the crime scene did not belong to the murderer?

_____
_____
_____

Name _____ Class _____ Date _____

DNA Whodunit *continued*

3. Interpreting Information Imagine you are on a jury and that DNA fingerprinting evidence is introduced. Explain how you would regard such evidence.

_____
_____
_____
_____
_____
_____

Extensions

1. **Research and Communications** Look through newspapers and news magazines to find articles about actual court cases in which DNA fingerprinting was used to determine the innocence or guilt of a suspect in a crime. Share the articles with your classmates.

2. **Research and Communications** Do library research or search the Internet to find out more information about restriction enzymes and what role they play in bacteria.

Name _____ Class _____ Date _____

Skills Practice Lab

BIOTECHNOLOGY

Introduction to Agarose Gel Electrophoresis

Gel electrophoresis is a process that is used to separate mixtures of electrically charged molecules, such as DNA and proteins, on the basis of their size and electrical charge. The process involves passing an electric current through a gel, which is a slab made of a jellylike substance. DNA is usually separated on gels made of *agarose*, a sugar that comes from certain types of marine algae.

During gel electrophoresis, each sample to be tested is placed in a depression called a *well*. An electric current applied across the gel causes one end of the gel to become negative and the other end to become positive. The electric current causes the samples to migrate through small holes, or pores, in the gel. Molecules that have a negative charge migrate toward the positive electrode. Molecules with a positive charge migrate toward the negative electrode. Small molecules move more easily through the pores in a gel than do large molecules. Therefore, smaller molecules move farther and at a faster rate than larger molecules. After gel electrophoresis, the largest molecules are found closest to their wells, while the smallest molecules are found the farthest away.

During gel electrophoresis, dyes are used to track the mobility of unstained DNA and indicate when the gel is finished running. In this lab, you will use dyes to explore the process of agarose gel electrophesis. You will use gel electrophoresis to observe the mobility of several dyes and to separate a dye mixture. You will then use your results to determine which dye(s) is/are missing from your unknown dye mixture.

OBJECTIVES

Demonstrate the separation of molecules based on size.

Demonstrate the separation of molecules based on charge.

Determine which component is missing in a mixture of several dyes.

MATERIALS

- agarose gel (2%)
- batteries connected in series, 9-V (5)
- beaker, 250 mL
- electrophoresis system, battery powered
- graduated cylinder, 250 mL
- lab apron
- micropipet, 10 μL
- micropipet tips (10)
- microtube of bromophenol blue (2)
- microtube of cystal violet
- microtube of orange G
- microtube of methyl green
- microtube of xylene cyanol
- microtube rack
- metric ruler (15 cm)
- safety goggles
- TBE running buffer, 200 mL (1×)

Copyright © by Holt, Rinehart and Winston. All rights reserved.

Holt Biology — How Proteins Are Made

Name _____ Class _____ Date _____

Introduction to Agarose Gel Electrophoresis *continued*

Procedure
PART 1: PIPETTING PRACTICE

1. Put on safety goggles and a lab apron.
2. Obtain from your teacher two microcentrifuge tubes—one that contains a colored solution and one that is empty.
3. Place a clean tip on the pipet.
4. Press the top button-like plunger down to the first stop. Immerse the tip into the tube that contains the sample.
5. *Slowly* release the plunger to draw the sample into the tip.
6. Move the pipet tip to the empty tube.
7. Expel the sample by pressing the plunger to the first stop, and then expel the remaining sample by pressing the plunger to the second stop. Do not release the plunger until the tip is out of the solution.
8. Press the ejector button to discard the tip.
9. Practice pipetting the solution between the two tubes until you are confident in your ability to pipet.

PART 2: GEL ELECTROPHORESIS OF DYES

10. Obtain dye samples from your teacher.
11. Place a new tip on the end of your micropipet.
12. Open the microtube containing bromophenol blue, and remove 10 μL of the dye.
13. Carefully pipet the solution into the well in Lane 1 of an agarose gel. To do this, place both elbows on the lab table, lean over the gel, and slowly lower the micropipet tip into the opening of the well before depressing the plunger. *Note: Do not jab the micropipet tip through the bottom of the well.*
14. Using a new micropipet tip for each tube, repeat steps 12 and 13 for each of the remaining samples. Place each dye in a well, according to the following lane assignments:

    Lane 2   Crystal violet   Lane 5   Xylene cyanol
    Lane 3   Orange G   Lane 6   Unknown Dye mixture
    Lane 4   Methyl green

Introduction to Agarose Gel Electrophoresis *continued*

FIGURE 1 SETTING UP THE ELECTROPHORESIS APPARATUS

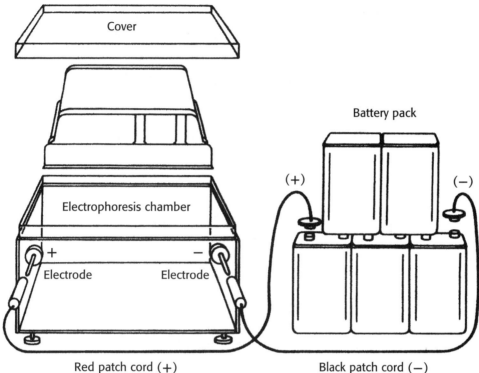

15. Carefully place the agarose gel (still in a gel-casting tray) in the electrophoresis chamber of an electrophoresis apparatus, as shown in **Figure 1.**

16. Slowly pour approximately 200 mL of 1×TBE running buffer into a beaker.

17. *Gently* and *slowly* pour the running buffer from the beaker into one side of the electrophoresis chamber until the gel is completely covered (approximately 1 to 2 mm above the top surface of the gel). *Note: Be careful not to overfill the chamber with buffer.*

18. Place the cover on the electrophoresis chamber. Wipe off any spills around the electrophoretic apparatus before doing the next step.

19. Connect five 9-V alkaline batteries as shown in **Figure 1. CAUTION: Do not touch both ends of the patch cords or both terminals on the battery pack at the same time.**

20. Connect the red (positive) patch cord to the red terminal on the chamber and the red terminal on the battery pack. Follow the same procedure with the black (negative) patch cord and the black terminals.

21. Observe the migration of the samples along the gel toward the red (positive) electrode and toward the black (negative) electrode.

22. Disconnect the battery pack when the dye bands in Lane 6 are fully separated and when one of the bands is near the end of the gel.

23. Remove the cover from the electrophoresis chamber.

Name _____ Class _____ Date _____

Introduction to Agarose Gel Electrophoresis *continued*

24. Lift the gel tray (containing the gel) from the chamber onto a sheet of paper towel. Notch one side of the gel so that you can identify the lanes. *Note: The dyes are not fixed on the gel and over time (several hours) may diffuse into the gel, making the bands less distinct. Complete the next step on the same day.*

25. Use a metric ruler to measure the distance of the dye bands in Lane 6 (in mm) from each of the six sample wells. *Note: Be sure to measure from the center of the well to the center of the band.*

26. Draw the dye bands in each gel lane. Make your drawing in the gel diagram shown in **Figure 2.**

FIGURE 2 MOBILITY OF DYES

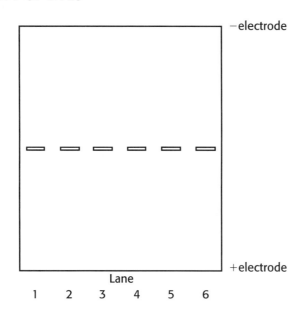

27. Label the dye bands in Lane 6 of your diagram alphabetically in order from top to bottom. Dye A will be closest to the negative pole.

28. In **Table 1,** record the migration of each sample toward (+) or away from (−) the positive pole. Then record the molecular charge (+ or −) of the dye in each band. Also record migration distance (in mm) for each band imaged on the gel. *Note: Measure from the wells in the center of the gel to each dye band*

29. Dispose of your materials according to the directions from your teacher.

30. Clean up your work area and wash your hands before leaving the lab.

Name _____ Class _____ Date _____

Introduction to Agarose Gel Electrophoresis *continued*

TABLE 1 EXPERIMENTAL DATA

	Migration direction [(+) or (−)]	Molecular charge (+ or −)	Migration distance (mm)
Lane 1			
Lane 2			
Lane 3			
Lane 4			
Lane 5			
Lane 6			
Dye A			
Dye B			
Dye C			
Dye D			
Dye E			

Analysis

1. Examining Data Which dye migrated the farthest distance? Which dye migrated the shortest distance?

_____

_____

2. Analyzing Results Which dye has the largest molecules? Which dye has the smallest molecules? Explain your answer.

_____

_____

_____

_____

Name _____ Class _____ Date _____

Introduction to Agarose Gel Electrophoresis *continued*

   3. **Explaining Events** Why do certain dyes migrate toward the cathode and others toward the anode?

   _____
   _____
   _____
   _____

Conclusions

   1. **Drawing Conclusions** What dyes make up the mixture of dyes that you loaded into Lane 6? Identify each dye in the mixture.

   _____
   _____
   _____
   _____

   2. **Drawing Conclusions** Which dye(s) was/were missing from your unknown mixture? Explain how you know this?

   _____
   _____
   _____
   _____

   3. **Interpreting Information** What characteristics of molecules allows them to be separated by gel electrophoresis?

   _____
   _____

Extensions

   1. **Designing Experiments** Design an experiment in which you determine what size DNA fragment each dye corresponds to. *Hint: Run a DNA ladder in one lane of an agarose gel and a dye mixture in another lane.*

   2. **Research and Communications** Research polyacrylamide gel electrophoresis. Find out how polyacrylamide gel electrophoresis is different from agarose gel electrophoresis and what applications researchers use it for.

TEACHER RESOURCE PAGE

Name _____ Class _____ Date _____

Quick Lab

DATASHEET FOR IN-TEXT LAB

Modeling Transcription

You can use paper and pens to model the process of transcription.

MATERIALS
- paper
- scissors
- pens or pencils (two colors)
- tape

Procedure

1. Cut a sheet of paper into 36 squares, each about 2.5 × 2.5 cm (1 × 1 in.) in size.

2. To make one side of your DNA model, line up 12 squares in a column. Using one color, randomly label each square with one of the following letters: A, C, G, or T. Each square represents a DNA nucleotide. Use tape to keep the squares in a column.

3. To make the second side of your DNA model, line up 12 squares next to the first column. Use the same color you used in step 2 to label each square with the complementary DNA nucleotide. Tape the squares together in a column.

4. Separate the two columns. The remaining 12 squares represent RNA nucleotides. Use a different color to "transcribe" one of the DNA strands.

Analysis

1. **Propose** a reason for using different colors for the DNA and RNA "nucleotides."

   Two colors represent the two different molecules.

2. **Predict** how a change in the sequence of nucleotides in a DNA molecule would affect the mRNA transcribed from the DNA molecule.

   The mRNA sequence would not be the same as the one constructed in the activity.

Copyright © by Holt, Rinehart and Winston. All rights reserved.

Holt Biology — How Proteins Are Made

Name _____ Class _____ Date _____

Modeling Transcription continued

3. Critical Thinking

Applying Information Use your model to test your prediction. Describe your results.

Their second mRNA is different from the first mRNA.

Decoding the Genetic Code

Data Lab

DATASHEET FOR IN-TEXT LAB

Background

Keratin is one of the proteins in hair. The gene for keratin is transcribed and translated by certain skin cells. The series of letters below represents the sequence of nucleotides in a portion of an mRNA molecule transcribed from the gene for keratin. This mRNA strand and the genetic code in the table below can be used to determine some of the amino acids in keratin.

mRNA Segment

U C U C G U G A A U U U U C C

First base	Second base				Third base
	U	**C**	**A**	**G**	
U	UUU ⎤ Phenylalanine UUC ⎦ UUA ⎤ Leucine UUG ⎦	UCU ⎤ UCC ⎥ Serine UCA ⎥ UCG ⎦	UAU ⎤ Tyrosine UAC ⎦ UAA ⎤ Stop UAG ⎦	UGU ⎤ Cysteine UGC ⎦ UGA – Stop UGG – Tryptophan	U C A G
C	CUU ⎤ CUC ⎥ Leucine CUA ⎥ CUG ⎦	CCU ⎤ CCC ⎥ Proline CCA ⎥ CCG ⎦	CAU ⎤ Histidine CAC ⎦ CAA ⎤ Glutamine CAG ⎦	CGU ⎤ CGC ⎥ Arginine CGA ⎥ CGG ⎦	U C A G
A	AUU ⎤ AUC ⎥ Isoleucine AUA ⎦ AUG – Start	ACU ⎤ ACC ⎥ Threonine ACA ⎥ ACG ⎦	AAU ⎤ Asparagine AAC ⎦ AAA ⎤ Lysine AAG ⎦	AGU ⎤ Serine AGC ⎦ AGA ⎤ Arginine AGG ⎦	U C A G
G	GUU ⎤ GUC ⎥ Valine GUA ⎥ GUG ⎦	GCU ⎤ GCC ⎥ Alanine GCA ⎥ GCG ⎦	GAU ⎤ Aspartic acid GAC ⎦ GAA ⎤ Glutamic acid GAG ⎦	GGU ⎤ GGC ⎥ Glycine GGA ⎥ GGG ⎦	U C A G

Analysis

1. **Determine** the sequence of amino acids that will result from the translation of the segment of mRNA above.

    Serine-arginine-glutamic acid-phenylalanine-serine

2. **Determine** the anticodon of each tRNA molecule that will bind to this mRNA segment.

    AGA, GCA, CUU, AAA, AGG

Name _____ Class _____ Date _____

Decoding the Genetic Code continued

3. Critical Thinking
   Recognizing Patterns Determine the sequence of nucleotides in the segment of DNA from which the mRNA strand on the previous page was transcribed.

   AGAGCACTTAAAAGG

4. Critical Thinking
   Recognizing Patterns Determine the sequence of nucleotides in the segment of DNA that is complementary to the DNA segment described in item 3.

   TCTCGTGAATTTTCC

Name _____ Class _____ Date _____

Quick Lab

DATASHEET FOR IN-TEXT LAB

Modeling Introns and Exons

You can use masking tape to represent introns and exons.

MATERIALS
- masking tape
- pens or pencils (two colors)
- metric ruler
- scissors

Procedure

1. Place a 15–20 cm strip of masking tape on your desk. The tape represents a gene.

2. Use two colors to write the words *appropriately joined* on the tape exactly as shown in the example. Space the letters so that they take up the entire length of the strip of tape. The segments in one color represent introns; those in the other color represent exons.

    appropriately joined

3. Lift the tape. Working from left to right, cut apart the groups of letters written in the same color. Stick the pieces of tape to your desk as you cut them, making two strips according to color and joining the pieces in their original order.

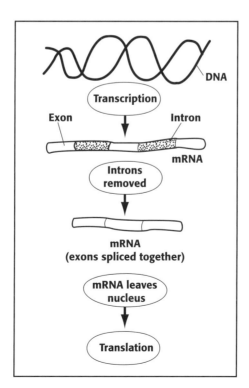

Name _____ Class _____ Date _____

Modeling Introns and Exons continued

Analysis

1. **Determine** from the resulting two strips which strip is made of "introns" and which is made of "exons."

   The strip with the letters *apprialyjoed* represents the introns. The strip with the letters that spell *protein* represent the exons.

2. **Critical Thinking**
   Predicting Outcomes Predict what might happen to a protein if an intron were not removed.

   Answers will vary. Because the function of a protein is ultimately a result of its amino acid sequence, a protein with additional amino acids will most likely not function.

TEACHER RESOURCE PAGE

Name _____ Class _____ Date _____

Exploration Lab

DATASHEET FOR IN-TEXT LAB

Modeling Protein Synthesis

SKILLS
- Modeling
- Using scientific methods

OBJECTIVES
- **Compare** and **Contrast** the structure and function of DNA and RNA.
- **Model** protein synthesis.
- **Demonstrate** how a mutation can affect a protein.

MATERIALS
- masking tape
- plastic soda-straw pieces of one color
- plastic soda-straw pieces of a different color
- paper clips
- pushpins of five different colors
- marking pens of the same colors as the pushpins
- 3 × 5 in. note cards
- oval-shaped card
- transparent tape

Before You Begin

The nature of a **protein** is determined by the sequence of amino acids in its structure. During **protein synthesis**, the sequence of nitrogen bases in an **mRNA** molecule is used to assemble **amino acids** into a protein chain.

A mutation is a change in the nitrogen-base sequence of DNA. Many mutations lead to altered or defective proteins. For example, the genetic blood disorder **sickle cell anemia** is caused by a mutation in the gene for **hemoglobin**.

In this lab, you will build models that will help you understand how protein synthesis occurs. You can also use the models to explore how a mutation affects a protein.

1. Write a definition for each boldface term in the paragraph above and for each of the following terms: transcription, translation, tRNA, ribosome, codon, anticodon. Use a separate sheet of paper. **Answers appear in the TE for this lab.**

Name _____ Class _____ Date _____

Modeling Protein Synthesis *continued*

2. Describe three differences between DNA and RNA.

   RNA is single stranded, and DNA is double stranded. RNA nucleotides contain the sugar ribose and the nitrogen bases uracil, guanine, cytosine, or adenine. DNA nucleotides contain the sugar deoxyribose and the nitrogen bases thymine, guanine, cytosine, or adenine.

3. Based on the objectives for this lab, write a question you would like to explore about protein synthesis.

   Answers will vary. For example: How does the cell's DNA determine the proteins made by a cell?

Procedure

PART A: DESIGN A MODEL

1. Work with the members of your lab group to design models of DNA, RNA, and a cell. Use the materials listed for this lab.

   > **You Choose**
   > As you design your models, decide the following:
   > a. what question you will explore
   > b. how to represent DNA nucleotides
   > c. how to represent RNA nucleotides
   > d. how to represent five different nitrogen bases
   > e. how to link (bond) nucleotides together
   > f. how to represent tRNA molecules with amino acids
   > g. how to represent the locations of DNA and ribosomes

2. Write out the plan for building your models. Have your teacher approve the plan before you begin building the models.

   Answers may vary. Students might use straw segments to model the sugar-phosphate backbone of DNA, different-colored pushpins to represent the four different nitrogen bases, paper clips to represent the bonds that hold the nucleotides in a chain, note cards to model tRNA molecules with amino acids attached, and oval-shaped cards to represent ribosomes.

Name _____ Class _____ Date _____

Modeling Protein Synthesis continued

3. Build the models your group designed. **CAUTION: Sharp or pointed objects may cause injury. Handle pushpins carefully.** Start your model of DNA with a strand of nucleotides that has the following sequence of nitrogen bases: TTTGGTCTCCTC.

PART B: MODEL PROTEIN SYNTHESIS

4. Use your models to demonstrate how transcription and translation occur. Draw and label the steps of each process on a separate sheet of paper. **Answers may vary. Check for students' understanding of each step in gene expression.**

5. Use your models to explore one of the questions written for step 3 of **Before You Begin.**

   Students' results will vary. Some students might produce a mutation and

   demonstrate how it affects protein synthesis. Some mutations cause the

   wrong amino acid to be inserted in the protein. Other mutations might insert

   a start or stop codon at an inappropriate place.

PART C: TEST HYPOTHESIS

Answer each of the following questions by writing a hypothesis. Use your models to test each hypothesis, and describe your results.

6. The DNA model you built for step 3 represents a portion of a gene for hemoglobin. Sickle cell anemia results from the substitution of an A for the T in the third codon of the nitrogen-base sequence given in step 3. How will this substitution affect a hemoglobin molecule?

   Hypotheses will vary. The substitution will change the amino acid sequence from

   lysine-proline-glutamic acid-glutamic acid to lysine-proline-valine-glutamic acid.

7. The addition of a nucleotide to a strand of DNA is a type of mutation called an *insertion*. What happens when an insertion occurs in the first codon in a DNA strand, before the DNA strand is transcribed?

   Hypotheses will vary. The codon triplets are shifted.

PART D: CLEANUP AND DISPOSAL

8. Dispose of damaged pushpins in the designated waste container.

9. Clean up your work area and all lab equipment. Return lab equipment to its proper place. Wash your hands thoroughly before you leave the lab and after you finish all work.

Name _____ Class _____ Date _____

Modeling Protein Synthesis continued

Analyze and Conclude

1. **Comparing Structures** How did the nitrogen-base sequence of the mRNA you made compare with that of the DNA it was transcribed from?

    They are complementary.

2. **Recognizing Relationships** How is the nitrogen-base sequence of a gene related to the structure of a protein?

    The nitrogen-base sequence of a gene contains a code that determines the amino-acid sequence of a protein.

3. **Recognizing Patterns** What is the relationship between the anticodon of a tRNA and the amino acid the tRNA carries?

    A tRNA with a particular anticodon always carries the same amino acid.

4. **Drawing Conclusions** How does a mutation in the gene for a protein affect the protein?

    A mutation can change the amino acid sequence of a protein, and thus may change its activity.

5. **Further Inquiry** Write a new question about protein synthesis that could be explored with your model.

    Example question: What happens if a mutation occurs that changes a codon to a stop codon?

TEACHER RESOURCE PAGE

Exploration Lab

DNA Whodunit

BIOTECHNOLOGY

Teacher Notes

TIME REQUIRED One 45-minute period

SKILLS ACQUIRED
Collecting data
Communicating
Constructing models
Identifying patterns
Inferring
Interpreting
Analyzing data

RATINGS
Easy ←— 1 2 3 4 —→ Hard

Teacher Prep–2
Student Setup–2
Concept Level–4
Cleanup–2

THE SCIENTIFIC METHOD

Make Observations Students make observations as they model enzyme digestion and gel electrophoresis.

Analyze the Results Students analyze results in Analysis questions 1 and 2.

Draw Conclusions Student draw conclusions in Conclusions question 1.

MATERIALS

Materials for this lab can be purchased as a kit from WARD'S. See the *Master Materials List* for ordering instructions.

Make copies of the Suspect/Victim DNA samples sheet. Cut apart the "Suspect" and "Victim" strips. Use a photocopy machine to make two additional strips from the DNA strip of the suspect that you choose to be the murderer and from the victim. Label these strips "crime scene sample 1" and "crime scene sample 2." For example, let's say you choose suspect 3 to be the murderer. Take a photocopy of the Suspect 3 DNA strip and cut out or blacken out the suspect's number and relabel as either "crime scene sample 1" or "crime scene sample 2." Do likewise with the victim's DNA strip. Assign one strip to each group in the class so that all strips are assigned at least once.

TIPS AND TRICKS

This lab works best in groups of two students.

Prepare an example of the gel electrophoresis page for students to see. You might want to prepare the gel electrophoresis pages for the entire class in advance.

The WARD'S kit has additional instructions on how to model the detection of the DNA through Southern blot analysis, which may be used to extend the scope of this lab.

TEACHER RESOURCE PAGE

DNA Whodunit continued

TEACHER'S GUIDE TO SUSPECT/VICTIM DNA AND RFLPS

Suspect 1 DNA
Ward II | Jan I | Jan I

Suspect 2 DNA
Ward II | Jan I

Suspect 3 DNA
Jan I | Ward II | Jan I

Suspect 4 DNA
Ward II | Jan I | Ward II

Suspect 5 DNA
Jan I | Ward II

Victim's DNA
Ward II | Jan I | Jan I

Key
- ⊕ = T
- ⊜ = A
- ◉ = G
- ● = C

Restriction Enzymes
Jan I
Ward II

Probe
X 38

Holt Biology — How Proteins Are Made

Name _____ Class _____ Date _____

Exploration Lab

BIOTECHNOLOGY

DNA Whodunit

In the early 1970s, scientists discovered that some bacteria have enzymes that are able to cut up DNA in a sequence-specific manner. These enzymes, now called *restriction enzymes*, recognize and bind to a specific short sequence of DNA, and then cut the DNA at specific sites within that sequence. Biologists found that they could use restriction enzymes to manipulate DNA. This ability formed the foundation for much of the biotechnology that exists today.

DNA fingerprinting is one important use of biotechnology. With the exception of identical twins, no two people have the same DNA sequence. Because each person has a DNA profile that is as unique as his or her fingerprints, DNA fingerprinting can be used to compare the DNA of different individuals.

In the first step of DNA fingerprinting, known and unknown samples are obtained and then digested, or cut into small fragments, by the same restriction enzyme. These short fragments are called *restriction fragment length polymorphisms (RFLPs)*. The next step in DNA fingerprinting is to separate the RFLPs by size. This is done with a technique called *gel electrophoresis*. The DNA is placed on a jellylike slab called a gel, and the gel is exposed to an electrical current. DNA has a negative electrical charge, so the RFLPs are attracted to the positive pole when an electric current is applied. Smaller fragments travel farther through the gel than longer ones. The length of a given DNA fragment can be determined by comparing its mobility on the gel with that of a sample containing DNA fragments of known sizes. The resulting pattern is unique for each individual.

In this lab, you will model experimental procedures involved in DNA fingerprinting and use your results to identify a hypothetical murderer.

OBJECTIVES

Use pop beads to model restriction enzyme digestion and agarose gel electrophoresis.

Evaluate the results of a model restriction enzyme digestion (DNA fingerprint).

Identify a hypothetical murderer by analyzing the simulated DNA fingerprints of suspects and DNA samples collected at the scene of the crime.

MATERIALS

- pop beads, blue (cytosine) (15)
- paper gel electrophoresis lane
- paper, legal size (8.5 × 14 in.)
- plastic connectors (hydrogen bonds) (30)
- pop beads, green (guanine) (15)
- pop beads, orange (thymine) (15)
- pop beads, red (phosphate) (60)
- pop beads, white, 5-hole (deoxyribose) (60)
- pop beads, yellow (adenine) (15)
- restriction enzyme card Jan I
- restriction enzyme card Ward II
- ruler

DNA Whodunit continued

Procedure

1. Read the following scenario.

    The police are investigating a murder. Blood stains of two different types were found at the murder scene. Based on other forensic evidence, the police have reason to believe that the murderer was wounded at the time of the murder. The police currently have five suspects for the murder. You have been provided with the DNA from a blood sample of one of the five suspects, or the DNA from one of the two blood stains found at the crime scene.

2. Assemble the DNA you were assigned with pop beads, using the DNA strip given to your group as a blueprint. Use **Figure 1** to guide you in your assembly of your DNA "molecule." Be sure to assemble the beads in the precise pattern indicated, or your results will be incorrect. The assembled chain represents your subject's DNA.

FIGURE 1 PATTERN FOR ASSEMBLY OF POP BEADS

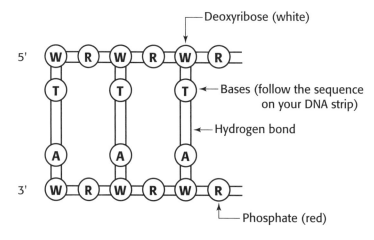

3. Place the DNA "molecule" you have just assembled on your work area so that the 5' end is on the top left side, as shown below. Be sure that the three orange beads (thymine) are in the following position on your work area:

    ```
    5'                      3'
        TTT, etc...............G
        AAA, etc...............C
    3'                      5'
    ```

Note: From this point on, it is important to keep the beads in this orientation. Do not allow the chain to be turned upside down or rotated. The 5' TTT end should always be on the top left of the molecule. If your chain is accidentally turned upside down, refer to your DNA strip to obtain the correct orientation.

TEACHER RESOURCE PAGE

Name _____ Class _____ Date _____

DNA Whodunit continued

4. Use the model restriction enzymes Jan I and Ward II to chop up the DNA. Look at your two Restriction Enzyme Cards; they look like the cards in **Figure 2.** These "enzymes" will make cuts in the DNA in the manner indicated by the dotted lines.

FIGURE 2 RESTRICTION ENZYME CARDS

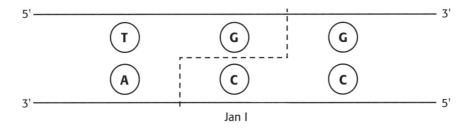
Jan I

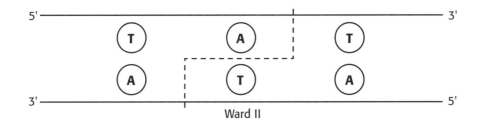
Ward II

5. Place Restriction Enzyme Card Jan I on top of the left side of the DNA chain so that its label is right side up.

6. Move the card along the surface of the DNA until you match the precise sequence shown on the card. When you reach a sequence that matches the card, stop and break the beads apart in the manner indicated by the dotted lines.

7. Move the enzyme card until you reach the right end of the DNA. Double check the sequence with the enzyme card to ensure that you have made all the possible cuts.

8. Repeat the procedure on the remaining DNA fragments using the restriction enzyme card Ward II. Be sure to keep the DNA fragments in the orientation described above (5' orange thymine beads on the top left) throughout this exercise. In reality, the fragments created in steps 6 and 7 might be thousands of base pairs long.

9. Create a gel electrophoresis area out of a legal size (8.5 by 14 in.) sheet of paper. On the left side of the paper, use a ruler to mark one inch increments from the bottom of the paper to the top of the paper. Starting at the bottom mark, label each mark from "0" (for the bottom mark) through '24" (for the top mark).

Copyright © by Holt, Rinehart and Winston. All rights reserved.

Holt Biology — How Proteins Are Made

Name _____ Class _____ Date _____

DNA Whodunit continued

10. Write a plus sign (+) at the bottom of the page and a minus sign (−) at the top of the page. Label the *Y*-axis (left hand margin) "Length of RFLPs (number of nucleotides."

11. Place the RFLPs at the negative pole of the gel electrophoresis page, taking care to retain the proper 5' to 3' orientation. Remember, DNA has a negative electrical charge, so the RFLPs are attracted to the positive end of the gel/page when an electric current is applied.

12. Simulate separating the RFLPs by electrophoresis by sliding your RFLPs along the gel/page. Shorter fragments are lighter and move farther than longer fragments. To determine the final position of each RFLP, count the number of nucleotides on the longest side of each fragment. Place each measured RFLP next to its corresponding length marked on the gel/page.

13. In the nine gel electrophoresis lanes in **Figure 3,** sketch dark bands at the correct positions in the gel lane reserved for your sample. Also, record the position of your bands on the seven lanes your teacher has provided for class data (on the blackboard).

14. Obtain the banding patterns for each of the other DNA samples by copying them from the blackboard after each team has recorded their data.

FIGURE 3 GEL ELECTROPHORESIS LANES

Victim's Blood	Suspect 1	Suspect 2	Suspect 3	Suspect 4	Suspect 5	Crime Scene Sample 1	Crime Scene Sample 2
22	22	22	22	22	22	22	22
20	20	20	20	20	20	20	20
18	18	18	18	18	18	18	18
16	16	16	16	16	16	16	16
14	14	14	14	14	14	14	14
12	12	12	12	12	12	12	12
10	10	10	10	10	10	10	10
8	8	8	8	8	8	8	8
6	6	6	6	6	6	6	6
4	4	4	4	4	4	4	4
2	2	2	2	2	2	2	2

Copyright © by Holt, Rinehart and Winston. All rights reserved.

Holt Biology   How Proteins Are Made

Name _____ Class _____ Date _____

DNA Whodunit *continued*

Analysis

1. Examining Data Are the RFLPs of the other DNA samples the same length as yours? Explain why or why not.

Answers will vary, but students should indicate that the lengths for most are different because the DNA of each individual is unique. However, two sets of samples will be the same; one DNA sample from the crime scene should match the DNA from the victim's blood and the other DNA sample from the crime scene should match the DNA of one of the suspect's blood.

2. Identifying Relationships Explain the role that restriction enzymes and gel electrophoresis play in DNA fingerprinting.

Restriction enzymse cut the DNA at a specific sequence, producing RFLPs.

Gel electrophoresis separates the RFLPs by size.

Conclusions

1. Drawing Conclusions Based on class data, which of the suspects is probably the murderer? Explain.

The identity of the murderer will vary. The suspect who's DNA sample matches that of the blood left at the crime scene (that is not the victim's blood) is probably the murderer.

2. Interpreting Information How did you show that the other sample found at the crime scene did not belong to the murderer?

The DNA fingerprint showed that the DNA from the other blood sample found at the crime scene matched that of the victim.

TEACHER RESOURCE PAGE

Name _____ Class _____ Date _____

DNA Whodunit *continued*

3. Interpreting Information Imagine you are on a jury and that DNA fingerprinting evidence is introduced. Explain how you would regard such evidence.

Answers will vary. Accept all reasonable answers. In the case presented here, the confirmation of the suspects blood at the murder scene does not mean that the suspect murdered the victim. It is only evidence that the suspect was at the crime scene and left blood there. The DNA evidence must be evaluated in conjuction with all of the other evidence before a conclusion can be drawn.

Extensions

1. **Research and Communications** Look through newspapers and news magazines to find articles about actual court cases in which DNA fingerprinting was used to determine the innocence or guilt of a suspect in a crime. Share the articles with your classmates.

2. **Research and Communications** Do library research or search the Internet to find out more information about restriction enzymes and what role they play in bacteria.

TEACHER RESOURCE PAGE

Skills Practice Lab

BIOTECHNOLOGY

Introduction to Agarose Gel Electrophoresis

Teacher Notes

TIME REQUIRED One 45-minute period

SKILLS ACQUIRED
Collecting data
Experimenting
Identifying patterns
Interpreting
Measuring
Organizing and analyzing data
Predicting

RATINGS
Easy ← 1 2 3 4 → Hard

Teacher Prep–4
Student Setup–3
Concept Level–3
Cleanup–3

THE SCIENTIFIC METHOD

Make Observations Students observe the mobility of several dyes.

Analyze the Results Analysis question 2 and Conclusions questions 1 and 2 require students to analyze their results.

Draw Conclusions Conclusions questions 1 and 2 ask student to draw conclusions from data.

MATERIALS

Materials for this lab activity can be purchased as a kit from WARD'S. See the *Master Materials List* for ordering instructions.

Precast gels come with the WARD'S kit. To cast 2.0% agarose gels, refer to agarose product literature. Prepare 1× TBE running buffer according to the manufacturer's instructions.

If you do not have automatic pipets, you may use disposable fine-tipped transfer pipets (1.6 mL) to deliver samples.

SAFETY CAUTIONS

- Discuss all safety symbols and caution statements with students.

- Be sure that students know the correct procedure for making electrical connections. Students should be supervised at all times during this investigation.

- Do not come in personal contact with or allow metal or any conductive material to come in contact with the reservoir buffer or the electrophorectic cell while the battery/power supply is on.

- If using a power supply, be sure to follow manufacturer's directions and precautions for safe handling.

Copyright © by Holt, Rinehart and Winston. All rights reserved.

Holt Biology — How Proteins Are Made

Introduction to Agarose Gel Electrophoresis continued

DISPOSAL

Dyes can be flushed down the drain with copious amounts of water.

Dilute 1× TBE buffer in a ratio of 1 part solution to 20 parts water. Flush the diluted solution down the drain with running water.

Stained agarose gels and other materials used in this lab can be put in the trash.

TECHNIQUES TO DEMONSTRATE

Demonstrate how to use a micropipet to load dye samples into gel wells. Caution students to be very careful not to puncture the gel with the micropipet tips when loading dye samples.

TIPS AND TRICKS

This lab works best in groups of three to five students.

If using variable volume pipets, make sure all pipets are set to 10 μL.

Make extra tubes of bromophenol blue dye solution for students to practice pipetting with.

To control the delivery of small volumes with transfer pipets, gently squeeze the pipet stem, instead of the bulb.

For this lab only, be sure to position and align the gel comb in the center of the casting tray.

Prelabel student microtubes (set of 6 per student team) as follows:

Tube 1	Bromophenol blue	Tube 4	Methyl green
Tube 2	Crystal violet	Tube 5	Xylene cyanol
Tube 3	Orange G	Tube 6	Dye mixture

Using a dropper, add 4 to 5 drops of each dye to corresponding microtubes.

Create unknown dye mixtures by adding one or two drops of each dyes, except the ones you choose to omit, to the dye mixture tube. Be sure to record in your notes which dye(s) have been omitted for each class.

Gels may also be loaded *after* the 1× TBE running buffer has been added. Loading the gel "wet" decreases the likelihood that the dyes samples will be washed out of the wells, but makes it harder to see the wells and avoid puncturing the bottom of the wells when loading. To "wet" load the gel, place a strip of black or dark colored paper under the electrophoresis chamber so that it is under the wells, making the wells more visible. Add 1× TBE running buffer until it covers the gel by 2 to 3 mm. Use a micropipet to load 10 μL into each well. Do not overload the wells.

The dyes on the gel are not fixed, and over time (about 24 hours) may diffuse, making the bands less distinct. Students must take the measurements on the same day.

The gel prepared in this lab does not require staining because the colored dyes will bind onto the gel and form clearly visible bands.

TEACHER RESOURCE PAGE

Name _____ Class _____ Date _____

Skills Practice Lab

BIOTECHNOLOGY

Introduction to Agarose Gel Electrophoresis

Gel electrophoresis is a process that is used to separate mixtures of electrically charged molecules, such as DNA and proteins, on the basis of their size and electrical charge. The process involves passing an electric current through a gel, which is a slab made of a jellylike substance. DNA is usually separated on gels made of *agarose*, a sugar that comes from certain types of marine algae.

During gel electrophoresis, each sample to be tested is placed in a depression called a *well*. An electric current applied across the gel causes one end of the gel to become negative and the other end to become positive. The electric current causes the samples to migrate through small holes, or pores, in the gel. Molecules that have a negative charge migrate toward the positive electrode. Molecules with a positive charge migrate toward the negative electrode. Small molecules move more easily through the pores in a gel than do large molecules. Therefore, smaller molecules move farther and at a faster rate than larger molecules. After gel electrophoresis, the largest molecules are found closest to their wells, while the smallest molecules are found the farthest away.

During gel electrophoresis, dyes are used to track the mobility of unstained DNA and indicate when the gel is finished running. In this lab, you will use dyes to explore the process of agarose gel electrophesis. You will use gel electrophoresis to observe the mobility of several dyes and to separate a dye mixture. You will then use your results to determine which dye(s) is/are missing from your unknown dye mixture.

OBJECTIVES

Demonstrate the separation of molecules based on size.

Demonstrate the separation of molecules based on charge.

Determine which component is missing in a mixture of several dyes.

MATERIALS

- agarose gel (2%)
- batteries connected in series, 9-V (5)
- beaker, 250 mL
- electrophoresis system, battery powered
- graduated cylinder, 250 mL
- lab apron
- micropipet, 10 μL
- micropipet tips (10)
- microtube of bromophenol blue (2)
- microtube of cystal violet
- microtube of orange G
- microtube of methyl green
- microtube of xylene cyanol
- microtube rack
- metric ruler (15 cm)
- safety goggles
- TBE running buffer, 200 mL (1×)

Copyright © by Holt, Rinehart and Winston. All rights reserved.

Holt Biology How Proteins Are Made

Name _____ Class _____ Date _____

Introduction to Agarose Gel Electrophoresis *continued*

Procedure

PART 1: PIPETTING PRACTICE

1. Put on safety goggles and a lab apron.
2. Obtain from your teacher two microcentrifuge tubes—one that contains a colored solution and one that is empty.
3. Place a clean tip on the pipet.
4. Press the top button-like plunger down to the first stop. Immerse the tip into the tube that contains the sample.
5. *Slowly* release the plunger to draw the sample into the tip.
6. Move the pipet tip to the empty tube.
7. Expel the sample by pressing the plunger to the first stop, and then expel the remaining sample by pressing the plunger to the second stop. Do not release the plunger until the tip is out of the solution.
8. Press the ejector button to discard the tip.
9. Practice pipetting the solution between the two tubes until you are confident in your ability to pipet.

PART 2: GEL ELECTROPHORESIS OF DYES

10. Obtain dye samples from your teacher.
11. Place a new tip on the end of your micropipet.
12. Open the microtube containing bromophenol blue, and remove 10 μL of the dye.
13. Carefully pipet the solution into the well in Lane 1 of an agarose gel. To do this, place both elbows on the lab table, lean over the gel, and slowly lower the micropipet tip into the opening of the well before depressing the plunger. *Note: Do not jab the micropipet tip through the bottom of the well.*
14. Using a new micropipet tip for each tube, repeat steps 12 and 13 for each of the remaining samples. Place each dye in a well, according to the following lane assignments:

    | Lane 2 | Crystal violet | Lane 5 | Xylene cyanol |
    | Lane 3 | Orange G | Lane 6 | Unknown Dye mixture |
    | Lane 4 | Methyl green | | |

Introduction to Agarose Gel Electrophoresis *continued*

FIGURE 1 SETTING UP THE ELECTROPHORESIS APPARATUS

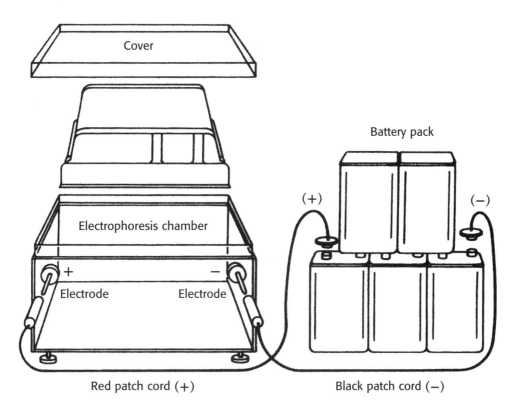

15. Carefully place the agarose gel (still in a gel-casting tray) in the electrophoresis chamber of an electrophoresis apparatus, as shown in **Figure 1.**

16. Slowly pour approximately 200 mL of 1×TBE running buffer into a beaker.

17. *Gently* and *slowly* pour the running buffer from the beaker into one side of the electrophoresis chamber until the gel is completely covered (approximately 1 to 2 mm above the top surface of the gel). *Note: Be careful not to overfill the chamber with buffer.*

18. Place the cover on the electrophoresis chamber. Wipe off any spills around the electrophoretic apparatus before doing the next step.

19. Connect five 9-V alkaline batteries as shown in **Figure 1. CAUTION: Do not touch both ends of the patch cords or both terminals on the battery pack at the same time.**

20. Connect the red (positive) patch cord to the red terminal on the chamber and the red terminal on the battery pack. Follow the same procedure with the black (negative) patch cord and the black terminals.

21. Observe the migration of the samples along the gel toward the red (positive) electrode and toward the black (negative) electrode.

22. Disconnect the battery pack when the dye bands in Lane 6 are fully separated and when one of the bands is near the end of the gel.

23. Remove the cover from the electrophoresis chamber.

Introduction to Agarose Gel Electrophoresis *continued*

24. Lift the gel tray (containing the gel) from the chamber onto a sheet of paper towel. Notch one side of the gel so that you can identify the lanes. *Note: The dyes are not fixed on the gel and over time (several hours) may diffuse into the gel, making the bands less distinct. Complete the next step on the same day.*

25. Use a metric ruler to measure the distance of the dye bands in Lane 6 (in mm) from each of the six sample wells. *Note: Be sure to measure from the center of the well to the center of the band.*

26. Draw the dye bands in each gel lane. Make your drawing in the gel diagram shown in **Figure 2**.

FIGURE 2 MOBILITY OF DYES

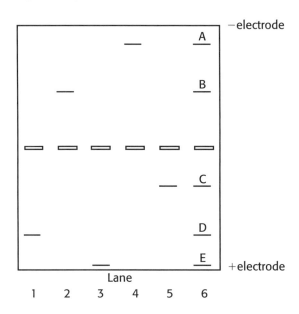

Each student group will have only three or four of the dye samples represented in Lane 6.

27. Label the dye bands in Lane 6 of your diagram alphabetically in order from top to bottom. Dye A will be closest to the negative pole.

28. In **Table 1,** record the migration of each sample toward (+) or away from (−) the positive pole. Then record the molecular charge (+ or −) of the dye in each band. Also record migration distance (in mm) for each band imaged on the gel. *Note: Measure from the wells in the center of the gel to each dye band*

29. Dispose of your materials according to the directions from your teacher.

30. Clean up your work area and wash your hands before leaving the lab.

Introduction to Agarose Gel Electrophoresis continued

TABLE 1 EXPERIMENTAL DATA

	Migration direction [(+) or (−)]	Molecular charge (+ or −)	Migration distance (mm)
Lane 1	(1)	2	23
Lane 2	(2)	1	15
Lane 3	(1)	2	31
Lane 4	(2)	1	28
Lane 5	(1)	2	10
Lane 6			
Dye A	(2)	1	28
Dye B	(2)	1	15
Dye C	(1)	2	10
Dye D	(1)	2	23
Dye E	(1)	2	31

Each student group will have only three or four of the dye samples represented in Lane 6.

Analysis

1. **Examining Data** Which dye migrated the farthest distance? Which dye migrated the shortest distance?

   Orange G, in Lane 3 migrated the farthest distance (31 mm). Xylene cyanol,

   in Lane 5 migrated the shortest distance (10 mm).

2. **Analyzing Results** Which dye has the largest molecules? Which dye has the smallest molecules? Explain your answer.

   Xylene cyanol, the dye in Lane 5, has the largest molecules, and Orange G,

   the dye in Lane 3 has the smallest. Smaller molecules can migrate farther

   through the gel and at a faster rate. Larger molecules are found nearer the

   point or origin.

Name _____ Class _____ Date _____

Introduction to Agarose Gel Electrophoresis *continued*

3. Explaining Events Why do certain dyes migrate toward the cathode and others toward the anode?

The positively charged molecules will migrate toward the cathode, and the negatively charged molecules will migrate toward the anode. As for the dyes used, those having a positive net charge will migrate toward the cathode, and those having a negative net charge will migrate toward the anode.

Conclusions

1. Drawing Conclusions What dyes make up the mixture of dyes that you loaded into Lane 6? Identify each dye in the mixture.

Answers will vary. Each unknown mixture should contain three or four of the dyes and be missing one or two of them.

2. Drawing Conclusions Which dye(s) was/were missing from your unknown mixture? Explain how you know this?

Answers will vary. Each unknown mixture should be missing one or two of the dyes. Students should be able to identify the missing dye(s) by comparing the color and mobility of the dyes in Lanes 1 through 5 with the color and mobility of the dyes in their unknown mixture (in Lane 6).

3. Interpreting Information What characteristics of molecules allows them to be separated by gel electrophoresis?

Size and charge allow molecules to be separated by gel electrophoresis.

Extensions

1. Designing Experiments Design an experiment in which you determine what size DNA fragment each dye corresponds to. *Hint: Run a DNA ladder in one lane of an agarose gel and a dye mixture in another lane.*

2. Research and Communications Research polyacrylamide gel electrophoresis. Find out how polyacrylamide gel electrophoresis is different from agarose gel electrophoresis and what applications researchers use it for.

Answer Key

Directed Reading

SECTION: FROM GENES TO PROTEINS

1. b
2. e
3. d
4. c
5. a
6. RNA polymerase
7. RNA
8. nucleus
9. Transcription makes RNA molecules, and DNA replication makes copies of DNA molecules. Also, in DNA replication, both strands of DNA are used as templates. In transcription, only one strand is used as a template.
10. Transcription begins at a gene's promoter, a specific sequence of DNA that acts as a "start" signal for a gene that is to be transcribed. Transcription ends at a sequence of bases that acts as a "stop" signal.
11. RNA is a type of nucleic acid. Messenger RNA is a form of RNA that carries the instructions for making a protein from a gene to the site of translation.
12. Codons are sequences of three nucleotides in an mRNA molecule that correspond to particular amino acids. The genetic code is the amino acids and "stop" and "start" signals coded for by mRNA codons.
13. 2
14. 6
15. 4
16. 3
17. 5
18. 1

SECTION: GENE REGULATION AND STRUCTURE

1. enzymes
2. promoter
3. operator
4. operon
5. *lac* operon
6. repressor
7. An enhancer is a sequence of DNA that can be bound by a transcription factor and thus influence transcription although it is located thousands of nucleotide bases away from the promoter.
8. A nuclear membrane separates transcription from translation in eukaryotes.
9. Gene expression is more complex in eukaryotes, and genes with related functions are often scattered on different chromosomes.
10. before, during, and after transcription and after mRNA leaves the nucleus or after the protein is functional
11. Introns are long segments of nucleotides in eukaryotic genes that have no coding information. Exons are the portions of a eukaryotic gene that are translated into proteins.
12. The mRNA that contains introns becomes smaller when enzymes cut out the introns and "stitch" the exons back together.
13. Introns might add evolutionary flexibility because each exon encodes a different part of the protein and cells can shuffle exons between genes, which makes new genes.
14. gametes
15. point
16. substitution
17. deletion
18. insertion

Active Reading

SECTION: FROM GENES TO PROTEINS

1. a. R e. R
   b. B f. B
   c. D g. D
   d. R h. B
2. c

TEACHER RESOURCE PAGE

SECTION: GENE REGULATION AND STRUCTURE

1. a change in the DNA of a gene
2. Because the mutation was passed to the individual's offspring, the original mutation must have occurred in a gamete.
3. Gene rearrangements occur when an entire gene is moved to a new location. Gene alterations involve a change in a gene itself, such as in the placement of the wrong amino acid during protein assembly.
4. An insertion is a mutation in which a sizable length of DNA is inserted into a gene.
5. When a deletion occurs, segments of a gene are lost. Thus, the information on the lost piece of DNA is no longer transferred to proteins. Also, a deletion will likely upset the triple groupings of the nucleotides following it, creating a new genetic message even in the part of gene that is not lost.
6. b

Vocabulary Review

ACROSS
1. RIBONUCLEIC
2. POLYMERASE
3. MESSENGER
4. GENE
6. TRANSCRIPTION
10. CODON
12. OPERATOR
14. INTRON

DOWN
1. REPRESSOR
5. EXON
6. TRANSLATION
7. ANTICODON
8. RIBOSOMAL
9. TRANSFER
11. GENETIC
13. OPERON

Science Skills

INTERPRETING TABLES
1. ACC
2. TTC
3. ATA
4. UGG
5. AAG
6. CUA
7. ACC
8. GAU
9. AUA
10. lysine
11. leucine
12. tyrosine
13. glutamic acid-arginine to aspartic acid-arginine
14. isoleucine-cysteine to isoleucine-tryptophan
15. cysteine-proline-proline to cysteine-phenylalanine-proline
16. glycine-leucine-threonine to glycine-stop
17. A frameshift mutation occurred. One of the G nucleotides and two C nucleotides have been deleted. The second codon is a stop codon, which will cause translation to end prematurely. The protein for that gene will be shortened and incomplete.

Concept Mapping

1. transcription
2. translation
3. genes or RNA polymerases
4. RNA polymerases or genes
5. RNA
6. amino acids

Critical Thinking

1. b
2. c
3. a
4. f
5. e
6. d
7. c
8. a
9. d
10. f
11. b
12. e
13. f, h
14. l, d
15. g, k
16. e, b
17. a, c
18. j, i
19. d
20. c
21. c
22. d
23. d

Quiz

SECTION: FROM GENES TO PROTEINS
1. a
2. b
3. a
4. a
5. d
6. c
7. e
8. a
9. b
10. d

SECTION: GENE REGULATION AND STRUCTURE
1. c
2. d
3. b
4. a
5. b
6. a
7. b
8. e
9. c
10. d

Test Prep Pretest

1. uracil
2. RNA polymerase
3. codons
4. A
5. operator
6. repressor
7. transcription factors
8. introns
9. exons
10. gene alterations
11. gene expression or protein synthesis
12. transcription
13. translation
14. d
15. c
16. a
17. b
18. d
19. RNA consists of a single strand of nucleotides instead of the two strands that form the DNA double helix. RNA nucleotides have the five-carbon sugar ribose rather than the sugar deoxyribose found in DNA nucleotides. RNA nucleotides have a nitrogen-containing base called uracil instead of the base thymine found in DNA nucleotides.
20. First mRNA binds to a ribosome. Then tRNAs carry amino acids to the ribosome according to the three-base codons on the mRNA. The amino acids are joined together to form a protein chain.
21. Messenger RNA (mRNA) is an RNA copy of a gene used as a blueprint for a protein. During translation, mRNA serves as a template for the assembly of amino acids. Transfer RNA (tRNA) molecules carry amino acids to the ribosome and act as interpreter molecules, translating mRNA sequences into amino acid sequences. Ribosomal RNA (rRNA) plays a structural role in ribosomes.
22. The *lac* operon consists of a cluster of genes that enables a bacterium to build the proteins needed for lactose metabolism only when lactose is present. Some of the genes determine whether or not other genes will be expressed; the other genes code for enzymes that break down lactose.
23. Eukaryotic cells contain more DNA than prokaryotic cells do. Although both prokaryotes and eukaryotes use regulatory proteins, eukaryotes use many more proteins than prokaryotes do, and the interactions are more complex. Instead of using operons, eukaryotic cells have genes with related functions scattered on different chromosomes.
24. Because of the existence of introns and exons, cells can shuffle exons between genes. Over time, this can result in new genes being made. For example, the 12 different hemoglobin genes in modern humans arose from 1 ancestral hemoglobin gene.
25. In point mutations called substitutions, one nucleotide in a gene is replaced with a different nucleotide. An insertion is the addition of one or more nucleotides to a gene. When a deletion occurs, one or more nucleotides are deleted from a gene.

Chapter Test (General)

1. d
2. d
3. a
4. d
5. c
6. b
7. b
8. c
9. a
10. c
11. b
12. d
13. f
14. h
15. a
16. c
17. e
18. g
19. b
20. d

Chapter Test (Advanced)

1. translation
2. anticodons
3. uracil
4. genetic code
5. repressor, off
6. RNA polymerase
7. eukaryotes
8. complementary
9. b
10. d
11. c
12. c
13. b
14. d
15. a
16. a
17. b
18. c
19. A gene's instructions for making a protein are transferred from DNA to mRNA during the process of transcription. In translation, tRNA, rRNA, and ribosomes use the instructions on the mRNA to put together the amino acids that make up the protein.
20. It suggests that all life-forms have a common evolutionary ancestor with a single genetic code.
21. When present in prokaryotic cells, lactose binds to the repressor protein and changes the protein's shape. The repressor prevents RNA polymerase from binding to the promoter. The change in shape releases the repressor. With the blocking effect eliminated, the transcription of genes that code for lactose-metabolizing enzymes proceeds.
22. Introns are long segments of nucleotides in eukaryotic genes with no coding information. After transcription, enzymes remove introns from the mRNA molecule before the mRNA is transcribed.
23. A gene alteration is a mutation that changes a gene. In a substitution, one nucleotide is replaced with another. Insertions and deletions involve the addition or omission, respectively, of one or more nucleotides. A point mutation involves a change in only one nucleotide.
24. Many eukaryotic genes are interrupted by segments of non-coding DNA called introns. The segments of DNA that are expressed are called exons. After transcription, exons are joined together and then translated.
25. Transcription factors are regulatory factors, which control mostly transcription in eukaryotes. In eukaryotes, an enhancer must be activated for a gene to be expressed. Transcription factors initiate transcription by binding to enhancers and to RNA polymerases.

TEACHER RESOURCE PAGE

Lesson Plan

Section: From Genes to Proteins

Pacing

Regular Schedule: with lab(s): 5 days   without lab(s): 3 days

Block Schedule: with lab(s): 2 1/2 days   without lab(s): 1 1/2 days

Objectives

1. Compare the structure of RNA with that of DNA.
2. Summarize the process of transcription.
3. Relate the role of codons to the sequence of amino acids that results after translation.
4. Outline the major steps of translation.
5. Discuss the evolutionary significance of the genetic code.

National Science Education Standards Covered

UNIFYING CONCEPTS AND PROCESSES

UCP1: Systems, order, and organization

UCP2: Evidence, models, and explanation

UCP3: Change, constancy, and measurement

UCP4: Evolution and equilibrium

UCP5: Form and function

SCIENCE AS INQUIRY

SI1: Abilities necessary to do scientific inquiry

SI2: Understandings about scientific inquiry

LIFE SCIENCE: THE CELL

LSCell1: Cells have particular structures that underlie their functions.

LSCell2: Most cell functions involve chemical reactions.

LSCell3: Cells store and use information to guide their functions.

LSCell4: Cell functions are regulated.

LIFE SCIENCE: THE MOLECULAR BASIS OF HEREDITY

LSGene1: In all organisms, the instructions for specifying the characteristics of the organisms are carried in DNA.

LSGene3: Changes in DNA (mutations) occur spontaneously at low rates.

Copyright © by Holt, Rinehart and Winston. All rights reserved.

Holt Biology — How Proteins Are Made

TEACHER RESOURCE PAGE

Lesson Plan *continued*

KEY

SE = Student Edition   TE = Teacher Edition
CRF = Chapter Resource File

Block 1

CHAPTER OPENER *(45 minutes)*

- **Quick Review,** SE. Students answer questions covered in previous sections of the textbook as preparation for the chapter content. (**GENERAL**)

- **Reading Activity,** SE. Students write a short list of the things that they already know about how proteins are made and things they would like to know about how proteins are made. (**GENERAL**)

- **Using the Figure,** TE. Students answer questions about the chapter opener photograph. (**GENERAL**)

- **Identifying Preconceptions,** TE. Point out that each cell (except sperm and eggs) contains the entire set of chromosomes, which includes two copies of every gene. (**GENERAL**)

Block 2

FOCUS *(5 minutes)*

- **Bellringer Transparency.** Use this transparency as students enter the classroom and find their seats. (**GENERAL**)

MOTIVATE *(10 minutes)*

- **Discussion/Question,** TE. Draw the structures of deoxyribose and ribose on the board. Ask the students to compare the two sugars. (**BASIC**)

TEACH *(30 minutes)*

- **Teaching Transparency, Section Outline.** Use this transparency to give students a framework for the information in this section. (**GENERAL**)

- **Inclusion Strategies,** TE. Students write each step of transcription and translation on numbered index cards, mix them up, and rearrange them in sequential order.

- **Teaching Transparency, Transcription: Making RNA.** Use this transparency to describe the events of transcription. Review the terms *nucleotide*, *RNA polymerase*, *template strand*, and *RNA*. Remind students that in RNA, the base uracil is substituted for thymine. (**GENERAL**)

- **Quick Lab,** Modeling Transcription, SE. Students make a paper model of transcription. (**GENERAL**)

- **Datasheets for In-Text Labs,** Modeling Transcription, CRF.

TEACHER RESOURCE PAGE

Lesson Plan continued

HOMEWORK

- **Teaching Tip**, Comparing Transcription and Replication, TE. Students make a graphic organizer to demonstrate the difference between transcription and DNA replication. A sample graphic organizer is provided in the TE. (**BASIC**)

- **Active Reading Worksheet**, From Genes to Proteins, CRF. Students read a passage related to the section topic and answer questions. (**GENERAL**)

- **Directed Reading Worksheet**, From Genes to Proteins, CRF. Students complete the exercises in this worksheet to help them understand the material as they read the section. (**BASIC**)

Block 3

TEACH *(30 minutes)*

- **Teaching Transparency, Codons in mRNA.** Use this transparency to discuss the genetic code. Familiarize students with the procedures for deciphering the genetic code key. Ask students why some codons can be considered synonyms. (**GENERAL**)

- **Teaching Transparency, Translation: Assembling Proteins.** Use this transparency to describe the structure of tRNA and the structure of a ribosome. Review the terms *codon, anticodon,* and *amino acid*. Describe how a new codon shifts into the A site. (**GENERAL**)

- **Data Lab,** Decoding the Genetic Code, SE. Students determine some of the amino acids in keratin from a model mRNA segment. (**GENERAL**)

- **Datasheets for In-Text Labs,** Decoding the Genetic Code, CRF.

- **Integrating Physics and Chemistry**, TE. Students name the enzyme that involved in the chemical digestion of lactose and identify the portion of the name that indicates it is an enzyme.

CLOSE *(15 minutes)*

- **Reteaching,** TE. Students play a game drawing vocabulary terms from a container and linking them conceptually. (**BASIC**)

- **Quiz,** TE. Students answer questions that review the section material. (**GENERAL**)

HOMEWORK

- **Alternative Assessment,** TE. Students create a colorful poster that compares and illustrates the functions of mRNA, tRNA, and rRNA. (**GENERAL**)

- **Inclusion Strategies**, TE. Students research the life of Barbara McClintock.

- **Section Review,** SE. Assign questions 1–6 for review, homework, or quiz. (**GENERAL**)

- **Quiz, CRF.** This quiz consists of ten multiple choice and matching questions that review the section's main concepts. (**BASIC**) **Also in Spanish.**

Copyright © by Holt, Rinehart and Winston. All rights reserved.

TEACHER RESOURCE PAGE

Lesson Plan *continued*

Optional Blocks

LABS *(90 minutes)*

- **Exploration Lab, DNA Whodunit, CRF.** Students model experimental procedures involved in DNA fingerprinting and use their results to identify a hypothetical murderer. (**GENERAL**)

- **Skills Practice Lab, Introduction to Agarose Gel Electrophoresis, CRF.** Students use gel electrophoresis to observe the mobility of several dyes and to separate a dye mixture. They then use their results to determine which dye(s) is/are missing from an unknown dye mixture. (**GENERAL**)

Other Resource Options

- **Internet Connect.** Students can research Internet sources about Genetic Code with SciLinks Code HX4089.

- **go.hrw.com.** For worksheets, videos, and other teaching aids related to this chapter, visit the HRW Web site and type in the keyword HX4 GNX.

- **Biology Interactive Tutor CD-ROM,** Unit 6 Gene Expression. Students watch animations and other visuals as the tutor explains gene expression. Students assess their learning with interactive activities.

- **CNN Science in the News, Video Segment 7 Gene Progress.** This video segment is accompanied by a **Critical Thinking Worksheet**.

- **CNN Student News.** Find the latest news, lesson plans, and activities related to important scientific events at **cnnstudentnews.com**.

Copyright © by Holt, Rinehart and Winston. All rights reserved.

Lesson Plan

Section: Gene Regulation and Structure

Pacing

Regular Schedule: with lab(s): 3 days   without lab(s): 2 days
Block Schedule: with lab(s): 1 1/2 days   without lab(s): 1 day

Objectives

1. Describe how the *lac* operon is turned on or off.
2. Summarize the role of transcription factors in regulating eukaryotic gene expression.
3. Describe how eukaryotic genes are organized.
4. Evaluate three ways that point mutations can alter genetic material.

National Science Education Standards Covered

UNIFYING CONCEPTS AND PROCESSES

UCP1: Systems, order, and organization

UCP2: Evidence, models, and explanation

UCP3: Change, constancy, and measurement

UCP4: Evolution and equilibrium

UCP5: Form and function

SCIENCE AS INQUIRY

SI1: Abilities necessary to do scientific inquiry

SI2: Understandings about scientific inquiry

LIFE SCIENCE: THE CELL

LSCell1: Cells have particular structures that underlie their functions.

LSCell3: Cells store and use information to guide their functions.

LSCell4: Cell functions are regulated.

LIFE SCIENCE: THE MOLECULAR BASIS OF HEREDITY

LSGene1: In all organisms, the instructions for specifying the characteristics of the organisms are carried in DNA.

LSGene2: Most of the cells in a human contain two copies of each of 22 different chromosomes. In addition, there is a pair of chromosomes that determine sex.

TEACHER RESOURCE PAGE

Lesson Plan *continued*

> **KEY**
> SE = Student Edition   TE = Teacher Edition
> CRF = Chapter Resource File

Block 4

FOCUS *(5 minutes)*

- **Bellringer Transparency.** Use this transparency as students enter the classroom and find their seats. **(GENERAL)**

MOTIVATE *(10 minutes)*

- **Demonstration**, TE. use a copy of the textbook to illustrate the percent of genetic information a typical cell uses. **(BASIC)**

TEACH *(30 minutes)*

- **Teaching Transparency, Section Outline.** Use this transparency to give students a framework for the information in this section. **(GENERAL)**

- **Demonstration**, TE. Using large colored–paper shapes to represent the components of the *lac* operon, demonstrate the process by which genes are turned on and off in the *lac* operon. **(GENERAL)**

- **Exploring Further**, Jumping Genes, SE. Students read this article and then discuss the role that transposons have on other genes. **(GENERAL)**

- **Teaching Transparency, Controlling Transcription in Eukaryotes.** Use this transparency to discuss the role of transcription factors in eukaryotes. **(GENERAL)**

HOMEWORK

- **Directed Reading Worksheet, Gene Regulation and Structure, CRF.** Students complete the exercises in this worksheet to help them understand the material as they read the section. **(BASIC)**

- **Active Reading Worksheet, Gene Regulation and Structure, CRF.** Students read a passage related to the section topic and answer questions. **(GENERAL)**

Block 5

TEACH *(30 minutes)*

- **Quick Lab,** Modeling Introns and Exons, SE. Students model introns and exons with masking tape. **(GENERAL)**

- **Datasheets for In-Text Labs, Modeling Introns and Exons, CRF.**

- **Teaching Transparency, Major Types of Mutations.** Use this transparency to illustrate the effects that mutations can have on genes. **(GENERAL)**

Copyright © by Holt, Rinehart and Winston. All rights reserved.

Holt Biology — How Proteins Are Made

TEACHER RESOURCE PAGE

Lesson Plan *continued*

- **Science Skills Worksheet, CRF.** Students interpret tables to decode the information in DNA and determine the effects of mutations on amino acid sequences. (**GENERAL**)

CLOSE *(15 minutes)*

- **Reteaching,** TE. Students work in groups to prepare a 5–minute presentation that summarizes one of the following topics: lac operon, eukaryotic gene expression, or mutations. Groups also prepare a four-question quiz about their topic. Have students answer the quiz in writing. (**BASIC**)

HOMEWORK

- **Teaching Tip**, Mutagens, TE. Pairs of students research a specific mutagen. (**ADVANCED**)
- **Quiz,** TE. Students answer questions that review the section material. (**GENERAL**)
- **Section Review,** SE. Assign questions 1–5 for review, homework, or quiz. (**GENERAL**)
- **Quiz, CRF.** This quiz consists of ten multiple choice and matching questions that review the section's main concepts. (**BASIC**) **Also in Spanish.**
- **Modified Worksheet, One-Stop Planner.** This worksheet has been specially modified to reach struggling students. (**BASIC**)
- **Critical Thinking Worksheet, CRF.** Students answer analogy-based questions that review the section's main concepts and vocabulary. (**ADVANCED**)

Optional Blocks

LAB *(45 minutes)*

- **Exploration Lab,** Modeling Protein Synthesis, SE. Students build models to explore protein synthesis and how mutations effect proteins. (**GENERAL**)
- **Datasheets for In-Text Labs,** Modeling Protein Synthesis, CRF.

Other Resource Options

- **Alternative Assessment,** TE. Students physically demonstrate the the operation of the lac operon and control of transcription in eukaryotes. (**GENERAL**)
- **Career,** Molecular Geneticist, TE. Discuss the job and importance of a molecular geneticist.
- **Supplemental Reading, A Feeling for the Organism, One-Stop Planner.** Students read the book and answer questions. (**ADVANCED**)
- **Internet Connect.** Students can research Internet sources about Genetic Disorders with SciLinks Code HX4091.
- **go.hrw.com.** For worksheets, videos, and other teaching aids related to this chapter, visit the HRW Web site and type in the keyword HX4 GNX.

Lesson Plan continued

- **Biology Interactive Tutor CD-ROM,** Unit 6 Gene Expression. Students watch animations and other visuals as the tutor explains gene expression. Students assess their learning with interactive activities.

- **CNN Science in the News, Video Segment 7 Gene Progress.** This video segment is accompanied by a **Critical Thinking Worksheet**.

Lesson Plan

End-of-Chapter Review and Assessment

Pacing

Regular Schedule: 2 days

Block Schedule: 1 day

KEY
SE = Student Edition TE = Teacher Edition
CRF = Chapter Resource File

Block 6

REVIEW *(45 minutes)*

- **Study Zone,** SE. Use the Study Zone to review the Key Concepts and Key Terms of the chapter and prepare students for the Performance Zone questions. **(GENERAL)**

- **Performance Zone,** SE. Assign questions to review the material for this chapter. Use the assignment guide to customize review for sections covered. **(GENERAL)**

- **Teaching Transparency, Concept Mapping.** Use this transparency to review the concept map for this chapter. **(GENERAL)**

Block 7

ASSESSMENT *(45 minutes)*

- **Chapter Test, How Proteins Are Made, CRF.** This test contains 20 multiple choice and matching questions keyed to the chapter's objectives. **(GENERAL) Also in Spanish.**

- **Chapter Test, How Proteins Are Made, CRF.** This test contains 25 questions of various formats, each keyed to the chapter's objectives. **(ADVANCED)**

- **Modified Chapter Test, One-Stop Planner.** This test has been specially modified to reach struggling students. **(BASIC)**

Other Resource Options

- **Vocabulary Review Worksheet, CRF.** Use this worksheet to review the chapter vocabulary. **(GENERAL) Also in Spanish.**

- **Test Prep Pretest, CRF.** Use this pretest to review the main content of the chapter. Each question is keyed to a section objective. **(GENERAL) Also in Spanish.**

- **Test Item Listing for ExamView® Test Generator, CRF.** Use the Test Item Listing to identify questions to use in a customized homework, quiz, or test.

- **ExamView® Test Generator, One-Stop Planner.** Create a customized homework, quiz, or test using the HRW Test Generator program.

Copyright © by Holt, Rinehart and Winston. All rights reserved.

TEST ITEM LISTING
How Proteins Are Made

TRUE/FALSE

1. ____ RNA nucleotides contain the five-carbon sugar ribose.
   Answer: True   Difficulty: I   Section: 1   Objective: 1

2. ____ Only DNA molecules contain the nitrogen base called uracil.
   Answer: False   Difficulty: I   Section: 1   Objective: 1

3. ____ During transcription, the information on a DNA molecule is "rewritten" into an mRNA molecule.
   Answer: True   Difficulty: I   Section: 1   Objective: 2

4. ____ A codon signifies either a specific amino acid or a stop signal.
   Answer: True   Difficulty: I   Section: 1   Objective: 3

5. ____ When a tRNA anticodon binds to an mRNA codon, the amino acid detaches from the tRNA molecule and attaches to the end of a growing protein chain.
   Answer: True   Difficulty: I   Section: 1   Objective: 3

6. ____ Only ribosomal RNA plays a role in translation.
   Answer: False   Difficulty: I   Section: 1   Objective: 3

7. ____ The genetic code is different in nearly all organisms.
   Answer: False   Difficulty: I   Section: 1   Objective: 4

8. ____ It has been discovered that each species of organism has its own unique genetic code for synthesis of its proteins.
   Answer: False   Difficulty: I   Section: 1   Objective: 4

9. ____ Cells regulate gene expression so that each gene will be transcribed only when it is needed.
   Answer: True   Difficulty: I   Section: 2   Objective: 1

10. ____ The operator portion of the *lac* operon controls RNA polymerase's access to lactose-metabolizing genes.
    Answer: True   Difficulty: I   Section: 2   Objective: 1

11. ____ Gene expression is prevented when a repressor binds to the group of genes involved in the same function.
    Answer: False   Difficulty: I   Section: 2   Objective: 1

12. ____ A repressor binds to the operator region when lactose is present.
    Answer: False   Difficulty: I   Section: 2   Objective: 1

13. ____ Repressor proteins are bound to the DNA in front of each gene, readily allowing transcription to take place as the RNA polymerase moves along that gene.
    Answer: False   Difficulty: I   Section: 2   Objective: 1

14. ____ An enhancer is a sequence of nucleotides that, when bound by transcription factors, aids in shielding the RNA polymerase binding site of a specific gene.
    Answer: False   Difficulty: I   Section: 2   Objective: 2

15. ____ When mRNA leaves the nucleus and enters the cytoplasm, it has a complete set of both introns and exons.
    Answer: False   Difficulty: I   Section: 2   Objective: 3

16. ____ Introns are deleted before a gene is transcribed from DNA into mRNA.
    Answer: False   Difficulty: I   Section: 2   Objective: 3

TEST ITEM LISTING, continued

17. ____ Introns are the portions of a gene that actually get translated into protein.
    Answer: False   Difficulty: I   Section: 2   Objective: 3

18. ____ A point mutation is the failure of a chromosome pair to separate during mitosis.
    Answer: False   Difficulty: I   Section: 2   Objective: 4

19. ____ Mutations that result from the substitution of one nitrogen base for another are called deletions.
    Answer: False   Difficulty: I   Section: 2   Objective: 4

MULTIPLE CHOICE

20. RNA differs from DNA in that RNA
    a. is single-stranded.
    b. contains a different sugar molecule.
    c. contains the nitrogen base uracil.
    d. All of the above
    Answer: D   Difficulty: I   Section: 1   Objective: 1

21. Which of the following is *not* found in DNA?
    a. adenine
    b. cytosine
    c. uracil
    d. None of the above
    Answer: C   Difficulty: I   Section: 1   Objective: 1

22. RNA is chemically similar to DNA except that its sugars have an additional oxygen atom, and the base thymine is replaced by a structurally similar base called
    a. uracil.
    b. alanine.
    c. cytosine.
    d. codon.
    Answer: A   Difficulty: I   Section: 1   Objective: 1

23. In RNA molecules, adenine is complementary to
    a. cytosine.
    b. guanine.
    c. thymine.
    d. uracil.
    Answer: D   Difficulty: I   Section: 1   Objective: 1

24. The function of rRNA is to
    a. synthesize DNA.
    b. synthesize mRNA.
    c. form ribosomes.
    d. transfer amino acids to ribosomes.
    Answer: C   Difficulty: I   Section: 1   Objective: 3

25. During transcription,
    a. proteins are synthesized.
    b. DNA is replicated.
    c. RNA is produced.
    d. translation occurs.
    Answer: C   Difficulty: I   Section: 1   Objective: 2

26. During transcription, the genetic information for making a protein is "rewritten" as a molecule of
    a. messenger RNA.
    b. ribosomal RNA.
    c. transfer RNA.
    d. translation RNA.
    Answer: A   Difficulty: I   Section: 1   Objective: 2

27. Transcription proceeds when RNA polymerase
    a. attaches to a ribosome.
    b. binds to a strand of DNA.
    c. binds to a strand of RNA.
    d. attaches to a promoter molecule.
    Answer: B   Difficulty: I   Section: 1   Objective: 2

28. Transcription is the process by which genetic information encoded in DNA is transferred to a(n)
    a. RNA molecule.
    b. DNA molecule.
    c. uracil molecule.
    d. transposon.
    Answer: A   Difficulty: I   Section: 1   Objective: 2

TEST ITEM LISTING, continued

29. Each nucleotide triplet in mRNA that specifies a particular amino acid is called a(n)
    a. mutagen.
    b. codon.
    c. anticodon.
    d. exon.

    Answer: B   Difficulty: I   Section: 1   Objective: 3

mRNA: CUCAAGUGCUUC

Genetic Code:

	U	C	A	G	
U	Phe	Ser	Tyr	Cys	U
	Phe	Ser	Tyr	Cys	C
	Leu	Ser	stop	stop	A
	Leu	Ser	stop	Trp	G
C	Leu	Pro	His	Arg	U
	Leu	Pro	His	Arg	C
	Leu	Pro	Gln	Arg	A
	Leu	Pro	Gln	Arg	G
A	Ile	Thr	Asn	Ser	U
	Ile	Thr	Asn	Ser	C
	Ile	Thr	Lys	Arg	A
	Met	Thr	Lys	Arg	G
G	Val	Ala	Asp	Gly	U
	Val	Ala	Asp	Gly	C
	Val	Ala	Glu	Gly	A
	Val	Ala	Glu	Gly	G

30. Refer to the illustration above. What is the portion of the protein molecule coded for by the piece of mRNA given?
    a. Ser—Tyr—Arg—Gly
    b. Val—Asp—Pro—His
    c. Leu—Lys—Cys—Phe
    d. Pro—Glu—Leu—Val

    Answer: C   Difficulty: II   Section: 1   Objective: 3

31. Refer to the illustration above. The anticodons for the codons in the mRNA given are
    a. GAG—UUC—ACG—AAG.
    b. GAG—TTC—ACG—AAG.
    c. CUC—GAA—CGU—CUU.
    d. CUU—CGU—GAA—CUC.

    Answer: A   Difficulty: II   Section: 1   Objective: 3

32. Refer to the illustration above. Which of the following would represent the strand of DNA from which the mRNA strand given was made?
    a. CUCAAGUGCUUC
    b. GAGUUCACGAAG
    c. GAGTTCACGAAG
    d. AGACCTGTAGGA

    Answer: C   Difficulty: II   Section: 1   Objective: 4

TEST ITEM LISTING, continued

mRNA codons	amino acid
UAU, UAC	tyrosine
CCU, CCC, CCA, CCG	proline
GAU, GAC	aspartic acid
AUU, AUC, AUA	isoleucine
UGU, UGC	cysteine

33. Refer to the illustration above. Suppose that you are given a protein containing the following sequence of amino acids: tyrosine, proline, aspartic acid, isoleucine, and cysteine. Use the portion of the genetic code given to determine which of the following contains a DNA sequence that codes for this amino acid sequence.
    a. AUGGGUCUAUAUACG
    b. ATGGGTCTATATACG
    c. GCAAACTCGCGCGTA
    d. ATAGGGCTTTAAACA
    Answer: B Difficulty: III Section: 1 Objective: 3

34. Each of the following is a type of RNA *except*
    a. carrier RNA.
    b. messenger RNA.
    c. ribosomal RNA.
    d. transfer RNA.
    Answer: A Difficulty: I Section: 1 Objective: 3

35. At the very beginning of translation, the first tRNA molecule
    a. binds to the ribosome's A site.
    b. attaches directly to the DNA codon.
    c. connects an amino acid to its anticodon.
    d. attaches to the P site of the ribosome.
    Answer: D Difficulty: I Section: 1 Objective: 3

36. A ribosome has
    a. one binding site for DNA.
    b. three binding sites used during translation.
    c. four binding sites for tRNA.
    d. no binding sites since the proteins must detach.
    Answer: B Difficulty: I Section: 1 Objective: 3

37. Transfer RNA
    a. carries an amino acid to its correct codon.
    b. synthesizes amino acids as they are needed.
    c. produces codons to match the correct anticodons.
    d. converts DNA into mRNA.
    Answer: A Difficulty: I Section: 1 Objective: 3

38. In order for translation to occur, mRNA must migrate to the
    a. ribosomes.
    b. *lac* operon.
    c. RNA polymerase.
    d. enhancer.
    Answer: A Difficulty: I Section: 1 Objective: 3

39. mRNA : nucleus ::
    a. nucleus : protein
    b. protein : cytoplasm
    c. nucleus : ribosomes
    d. protein : nucleus
    Answer: B Difficulty: II Section: 1 Objective: 3

TEST ITEM LISTING, continued

40. A site : tRNA ::
    a. codon : mRNA
    b. mRNA : amino acid
    c. tRNA : amino acid
    d. mRNA : P site

    Answer: C   Difficulty: II   Section: 1   Objective: 3

41. codon : mRNA ::
    a. P site : RNA molecules
    b. ribosome : DNA molecules
    c. DNA : protein
    d. anticodon : tRNA

    Answer: D   Difficulty: II   Section: 1   Objective: 3

42. During translation, the amino acid detaches from the transfer RNA molecule and attaches to the end of a growing protein chain when
    a. the ribosomal RNA anticodon binds to the messenger RNA codon.
    b. the transfer RNA anticodon binds to the messenger RNA codon.
    c. a "stop" codon is encountered.
    d. the protein chain sends a signal through the nerve cells to the brain.

    Answer: B   Difficulty: I   Section: 1   Objective: 3

43. In bacteria, a group of genes that code for functionally related enzymes, their promoter site, and the operator that controls them all function together as a(n)
    a. exon.
    b. intron.
    c. operon.
    d. ribosome.

    Answer: C   Difficulty: I   Section: 2   Objective: 1

44. The function of an operator is to
    a. regulate access of RNA polymerase to specific genes.
    b. turn on and off the molecules of tRNA.
    c. control the process of transcription within the nucleus.
    d. generate amino acids for protein synthesis.

    Answer: A   Difficulty: I   Section: 2   Objective: 1

45. Cells must control gene expression so that
    a. their genes will be expressed only when needed.
    b. their genes will always be expressed.
    c. their genes will never be expressed.
    d. genetic disorders can be corrected.

    Answer: A   Difficulty: I   Section: 2   Objective: 1

46. A repressor protein
    a. prevents DNA synthesis.
    b. blocks movement of RNA polymerase.
    c. attaches to ribosomes during translation.
    d. destroys amino acids before protein synthesis occurs.

    Answer: B   Difficulty: I   Section: 2   Objective: 1

47. The presence of a repressor protein prevents the action of what enzyme?
    a. DNA polymerase
    b. lactase
    c. RNA polymerase
    d. permease

    Answer: C   Difficulty: I   Section: 2   Objective: 1

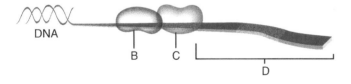

TEST ITEM LISTING, continued

48. Refer to the illustration above. To which portion of the *lac* operon does the repressor bind?
    a. regulator
    b. B
    c. C
    d. D
    Answer: C   Difficulty: II   Section: 2   Objective: 1

49. Refer to the illustration above. Where on the *lac* operon does transcription take place?
    a. regulator
    b. B
    c. C
    d. D
    Answer: D   Difficulty: II   Section: 2   Objective: 1

50. Where on the *lac* operon does a repressor molecule bind when lactose is absent?
    a. to the operator
    b. to the promoter
    c. to a structural gene
    d. to the regulator
    Answer: A   Difficulty: I   Section: 2   Objective: 1

51. The *lac* operon is shut off when
    a. lactose is present.
    b. lactose is absent.
    c. glucose is present.
    d. glucose is absent.
    Answer: B   Difficulty: I   Section: 2   Objective: 1

52. Transcription factors are
    a. enhancers.
    b. promoters.
    c. regulatory proteins.
    d. None of the above
    Answer: C   Difficulty: I   Section: 2   Objective: 2

53. The portions of DNA molecules that actually code for the production of proteins are called
    a. mutons.
    b. exons.
    c. introns.
    d. exposons.
    Answer: B   Difficulty: I   Section: 2   Objective: 3

54. The non-coding portions of DNA that are separated from the portions of DNA actually used during transcription are called
    a. mutons.
    b. exons.
    c. introns.
    d. exposons.
    Answer: C   Difficulty: I   Section: 2   Objective: 3

55. Many biologists believe that having the genes of eukaryotic cells interrupted by introns
    a. prevents the code from being copied.
    b. causes severely damaging mutations.
    c. ensures that replication occurs correctly.
    d. provides evolutionary flexibility.
    Answer: D   Difficulty: I   Section: 2   Objective: 3

56. Many thousands of proteins may have arisen from only a few thousand exons because
    a. an exon may be used by many different genes.
    b. there really is no difference between one protein and another.
    c. an exon does not actually code for any meaningful information.
    d. one gene can code for hundreds of different proteins.
    Answer: A   Difficulty: I   Section: 2   Objective: 3

TEST ITEM LISTING, continued

COMPLETION

57. The nitrogen-containing base that is only found in RNA is _____.
    Answer: uracil   Difficulty: I   Section: 1   Objective: 1

58. The enzyme responsible for making RNA is called _____.
    Answer: RNA polymerase   Difficulty: I   Section: 1   Objective: 2

59. A sequence of DNA at the beginning of a gene that signals RNA polymerase to begin transcription is called a(n) _____.
    Answer: promoter   Difficulty: I   Section: 1   Objective: 2

60. Messenger RNA is produced during the process of _____.
    Answer: transcription   Difficulty: I   Section: 1   Objective: 2

61. Transcription and translation are stages in the process of _____.
    Answer: gene expression   Difficulty: I   Section: 1   Objective: 2

62. The first stage of gene expression is called _____.
    Answer: transcription   Difficulty: I   Section: 1   Objective: 2

63. During translation, amino acids are brought to the ribosomes by molecules of _____.
    Answer: transfer RNA   Difficulty: I   Section: 1   Objective: 3

64. Nucleotide sequences of tRNA that are complementary to codons on mRNA are called _____.
    Answer: anticodons   Difficulty: I   Section: 1   Objective: 3

65. The sequence of three nucleotides that code for specific amino acids or stop signals in the synthesis of protein is called a(n) _____.
    Answer: codon   Difficulty: I   Section: 1   Objective: 3

66. The information contained in a molecule of messenger RNA is used to make protein during the process of _____.
    Answer: translation   Difficulty: I   Section: 1   Objective: 3

67. The form of ribonucleic acid that carries genetic information from the DNA to the ribosomes is _____.
    Answer: mRNA   Difficulty: I   Section: 1   Objective: 3

68. Cells must regulate gene expression so that genes will be _____ only when the proteins are needed.
    Answer: transcribed   Difficulty: I   Section: 2   Objective: 1

69. A cluster of genes in a bacterial cell that codes for proteins with related functions is called a(n) _____.
    Answer: operon   Difficulty: I   Section: 2   Objective: 1

70. A protein that prevents transcription by blocking the path of RNA polymerase along a molecule of DNA is called a(n) _____.
    Answer: repressor   Difficulty: I   Section: 2   Objective: 1

71. Transcription begins when an enzyme called _____ binds to the beginning of a gene on a region of DNA called a promoter.
    Answer: RNA polymerase   Difficulty: I   Section: 2   Objective: 2

TEST ITEM LISTING, *continued*

72. Nucleotide segments of a DNA molecule that make up genes and are actually expressed in the phenotype of an organism are called _____.
    Answer: exons Difficulty: I Section: 2 Objective: 3

73. Portions of genes that actually get translated into proteins are called _____.
    Answer: exons Difficulty: I Section: 2 Objective: 3

74. Mutations that change a gene are called gene _____.
    Answer: alterations Difficulty: I Section: 2 Objective: 4

ESSAY

75. Identify the three types of RNA and briefly describe the function of each.
    Answer:
    Three types of RNA are messenger RNA (mRNA), transfer RNA (tRNA), and ribosomal RNA (rRNA). Messenger RNA carries hereditary information from the DNA in the nucleus to the site of translation on the ribosomes. tRNA carries amino acids to the ribosomes for assembly into proteins. rRNA is a structural molecule, becoming part of the ribosomes upon which translation occurs.
    Difficulty: III Section: 1 Objective: 3

76. Genes control cellular activities through a two-step process known as gene expression. Name and discuss the significance of the two steps.
    Answer:
    Information encoded in DNA molecules undergoes transcription as RNA polymerase makes an mRNA molecule with nucleotides having a sequence that is complementary to that of one of the original DNA strands. The mRNA molecule leaves the nucleus and associates with a ribosome, where the second step, translation, occurs. Translation involves the synthesis of the amino acid sequence of a protein molecule by the combined action of mRNA, tRNA, and rRNA. The sequence of mRNA nucleotides determines the sequence of amino acids in the assembled protein.
    Difficulty: III Section: 1 Objective: 3

77. What is the evolutionary significance of the genetic code?
    Answer:
    The fact that the genetic code is the same in all organisms and is nearly universal suggests that all life-forms have a common evolutionary ancestor with a single genetic code.
    Difficulty: II Section: 1 Objective 4

78. Describe the physical structure of the *lac* operon.
    Answer:
    The *lac* operon consists of three segments. These include a promoter, an operator, and three lactose-metabolizing genes. In addition, a regulator gene lies close to the *lac* operon.
    Difficulty: III Section: 2 Objective: 1

79. In a mutant strain of *Escherichia coli,* lactose fails to bind to the repressor on the operator portion of the *lac* operon. What is likely to be the result of this failure?
    Answer:
    The failure of lactose to bind to and remove the repressor will prevent the *lac* operon from functioning. As a result, RNA polymerase will not transcribe the lactose-metabolizing genes of the *lac* operon, and the enzymes that normally break down lactose will not be produced.
    Difficulty: III Section: 2 Objective: 1

TEST ITEM LISTING, continued

80. In the *lac* operon, how does RNA polymerase affect the expression of the lactose-metabolizing genes, and how is the activity of RNA polymerase controlled?

    Answer:
    RNA polymerase is needed to transcribe the DNA code into mRNA. As long as the repressor is attached to the operon, the activity of the RNA polymerase is prevented. When lactose binds to and removes the repressor, the RNA polymerase can move to the lactose-metabolizing genes of the *lac* operon, and mRNA can be transcribed.

    Difficulty: III   Section: 2   Objective: 1

81. Describe two types of mutations and their effects.

    Answer:
    Gene rearrangements are mutations that move an entire gene to a new location. Changes in a gene's position often disrupt the gene's function because the gene is exposed to new regulatory controls in its new location. Gene alterations are changes in a gene itself. These include a change in one or many nucleotides in a gene. Point mutations, insertions, and deletions are examples of gene alterations. These types of mutations can disrupt a gene's function. (Students might also describe point mutations, insertions, deletions, or transpositions in response to this question.)

    Difficulty: III   Section: 2   Objective: 4